F. D. O'REILLY and P. I. McDONALD

THAILAND'S AGRICULTURE

GEOGRAPHY OF WORLD AGRICULTURE

12

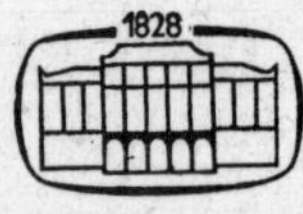

AKADÉMIAI KIADÓ · BUDAPEST 1983

F. D. O'REILLY and P. I. McDONALD

THAILAND'S AGRICULTURE

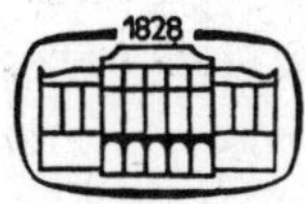

AKADÉMIAI KIADÓ · BUDAPEST 1983

English text revised by

PAUL A. COMPTON
The Queen's University of Belfast
Northern Ireland

ISBN 963 05 3360 X
HU-ISSN 0303-6634

CONTENTS

5

ACKNOWLEDGMENTS

We are indebted to many organizations in Thailand for their help, both direct and indirect, with this study: in particular the various ministries of the Royal Thai Government, the Mekong Secretariat and the Applied Scientific Research Council of Thailand.

We are grateful to the late Professor C. A. Fisher of the Department of Geography of the School of Oriental and African Studies of the University of London for his encouragement of this project at an early stage.

Thanks are due to the following Thai friends for generous help and encouragement: Khun Prapai Panit (R.I.D.), Khun Pittaya Hiranburana (R.I.D.), Khun Chitraporn Tulawattana (Mekong Office) and Khun Napaporn Ratitamkul (Advance Pharma).

We would also like to thank Mrs Anette Percy and Mrs Eileen Bradshaw for typing the manuscript and Mr Michael Stephens for drawing the maps and diagrams.

London, August 1979

F.D. O'Reilly
P.I. McDonald

KEY TO THAILAND'S PROVINCES

Metropolitan
1 Phra Nakhon
2 Thon Buri
3 Samut Prakan
4 Nonthaburi

Central Plain
4 Pathum Thani
6 Phra Nakhon Si Ayutthaya
7 Ang Thong
8 Sing Buri
9 Chai Nat
10 Uthai Thani
11 Suphan Buri
12 Nakhon Pathom
13 Samut Sakhon

West
14 Kanchanaburi
15 Ratchaburi
16 Phetchaburi
17 Samut Songkhram
18 Prachuap Khiri Khan

East
19 Nakhon Nayok
20 Prachin Buri
21 Chachoengsao
22 Chon Buri
23 Rayong
24 Chanthaburi
25 Trat

Upper Central Plain
26 Sara Buri
27 Lop Buri
28 Nakhon Sawan
29 Phetchabun
30 Phichit
31 Phitsanulok
32 Uttaradit
33 Sukhothai
34 Kamphaeng Phet
35 Tak

North
36 Phrae
37 Nan
38 Chiang Rai
39 Lampang
40 Lamphun
41 Chiang Mai
42 Mae Hong Son

North North East
43 Loei
44 Udon Thani
45 Nong Khai
46 Sakon Nakhon
47 Nakhon Phanom

Near North East
48 Khon Kaen
49 Chaiyaphum
50 Nakhon Ratchasima
51 Buri Ram

East North East
52 Maha Sarakham
53 Kalasin
54 Roi Et
55 Surin
56 Si Sa Ket
57 Ubon Ratchathani

South
58 Chumphon
59 Ranong
60 Phangnga
61 Phuket
62 Surat Thani
63 Krabi
64 Nakhon Si Thammarat
65 Trang
66 Phatthalung
67 Satun
68 Songkhla
69 Yala
70 Pattani
71 Narathiwat

CHAPTER ONE

PROBLEMS IN THE DETERMINATION
OF AGRICULTURAL REGIONS
IN DEVELOPING COUNTRIES

The determination of agricultural regions in the so-called developing countries of the Third World poses for the scholar special and peculiar problems, the magnitude of which varies from country to country. These problems do not normally arise in the study of agricultural regions in the developed industrial countries, be they capitalist or socialist.

The first problem derives from the inherently subsistence nature of agriculture in many developing countries. Some regions are entirely untouched by commercial agriculture. Other regions do produce cash crops for the national or world market, yet still deserve the epithet "subsistent", because the primary motivation of their farmers is the provision of sustenance for themselves and their families. Thus they practise a dual economy, cultivating on part of the farm their subsistence staples and on other parts their cash crops. The nature and the dynamism of the ratio between these two activities in terms of areal extent and inputs and outputs are crucial.

The subsistence element lends a monotony to the agriculture of a developing country, which militates against the determination of agricultural regions, although in a geographically or ethnically diverse nation the specific subsistence crops may vary. In Thailand though there is no part of the country which does not grow rice, the nation's staple. Indeed, in most parts of the country it is the dominant crop. In a sense, then, the whole of Thailand is a rice region. This fundamental fact should be borne in mind in the ensuing pages which are basically concerned with spatial variations in Thai agriculture.

A subsistence farmer does not, however, grow only a staple crop. He also endeavours to provide his family with a wide variety of vegetables, herbs and fruit. Thus in Thailand such garden or *suan* crops occur widely throughout the country, normally being cultivated in small plots around the house, and may easily, through their seeming insignificance, escape the attention of the agricultural census or land-use survey. This situation contrasts with the industrially developed countries where fruit and vegetables are often grown by specialist farmers in distinctive agricultural regions.

Third World agriculture is often considered to be conservative in contrast to the more dynamic agriculture of the developed countries. In practice, however, the desire to commercialize a basically subsistence agriculture often leads to rapid changes in crop complexes. New crops come and go with surprising and disheartening rapidity, the spread and decline of kenaf in North East Thailand being a classic case in point. Few farmers in the developed countries are required or are prepared to effect the radical changes in their crop systems that their less educated counterparts in the developing countries are constantly exhorted or pressured to do. Thus cash crop regions in the developing countries rarely have the stability of such regions in the developed countries. In delineating agricultural regions in Thailand the writers are acutely aware that they are describing a specific point in time and that the overall picture of cash crop regions may have altered, not only in details but also in fundamentals, in just a few years' time.

Multiple cropping and inter-cropping also make the researchers' task difficult. Double or treble cropping is possible in humid tropical areas where temperatures permit plant growth throughout the year, provided sufficient water is available. Double cropping of, for example, rice leads to no problem in land-use delineation, although it may lead the unwary scholar to underestimate the productive potential of the land. If, on the other hand, a given field carries in one year first a crop of rice and then a crop of water melons (as was observed in North East Thailand), this poses problems in the construction of land-use maps. Land-use surveys, especially if conducted by the use of aerial photography, increasingly popular, instead of trekking the fields may hopelessly fail to take account of the unseen double or treble cropping.

Inter-cropping is an ingenious device to maximize the use of a small farm holding, which makes economic and ecological sense whereby, in one field, the farmer grows a range of complementary crops simultaneously. Thus in a field of mixed millet and sorghum, a lower level may be occupied by groundnuts, whilst cowpeas ramble along at ground level (as is observed in West Africa). Thus there are distinct plant storeys. Tall plants protect lower ones from the direct sunlight, whilst ground-level plants protect the soil from the ravages of erosion. Differing maturing rates and harvest times even out the farm's labour requirements. However, inter-cropping does complicate the agricultural land-use map and of necessity the resultant agricultural regions.

A further complication is the practice of shifting cultivation. This is a rotation of fields rather than of crops. After a period of cultivating a particular farm the farmer allows his land to revert to the wild in order to regain its fertility, whilst he moves on to another site of natural or secondary vegetation, which he clears and upon which he re-establishes his farm. Sooner or later, depending upon the ratio between population and available land resources, the farmer must return to his original site. This system of cultivation means that the actual farm land of a given village may be several times greater than the observed farmed area. Similarly, what seems to the observer to be natural woodland or waste land, may simply be forest fallow. In Thailand, although shifting cultivation is not a dominant activity, as it is in some countries, it is important among the tribal groups of Northern Thailand, who occupy quite a large area and constitute a national problem out of all proportion to their relatively small population size. Just as the Hill Tribes escape the Population Census, so their system of agriculture goes unrecorded in the national Census of Agriculture.

Pastoral nomadism obviously cannot be fitted into an agricultural regionalization scheme at all. Pastoral nomadism, a problem in much of the savanna belt of Africa, is unknown in Thailand. Herbivores (cattle, water buffaloes and goats) are kept by the Thai farmer. However, there are no areas of the farm specifically set aside as pasture land for these beasts. Rather they graze the vegetation on road and canal sides as well as the voluntary vegetation on harvested fields. Thus a field of rice after harvest may serve as a pasture for cattle, which in turn provide manure *in situ*. This, yet another example of the multiple use of limited land resources, further complicates land-use studies.

In the ensuing pages the writers have relied primarily upon official Thai statistics, *faute de mieux*. The aforementioned problems of course faced the official compilers of these statistics, although it is not usual for officialdom either to admit such problems or to explain the methods used to circumvent them. Researchers in developing countries often cavil at the unavailability or unreliability of official statistics, but it must be remembered that Thai farmers are not as used to precision as their European counterparts; therefore, it is likely that statistics derived from questioning farmers are at best only fair approximations. Available relevant official statistics may not always be available for the same year, and the researcher has to compare for example rice figures for one year with rubber figures for another year and arrive at an agricultural regionalization from this compromise.

None of these problems should, however, deter the researcher from attempting to determine agricultural regions within developing countries. By doing so, the service he renders is a worthwhile one for in developing countries agricultural policies and plans are often formulated without an adequate data base, and the determination of agricultural regions can certainly aid the planning of agricultural development.

Is it, however, valid to compare agricultural regions in developing countries with those in the developed world? Is there any common ground between such studies in Thailand or Nigeria and those in Belgium or Czechoslovakia? Has a world map of agricultural regions any real meaning? Suffice it to say that the writers feel that such comparisons are of use and that the attempt to describe and regionalize world agriculture is worthwhile. We live in one shrinking, vulnerable world. Together we face the problem of depletion of resources, yet the resources of the earth are still inadequately known. Moreover international organizations are attempting increasingly to plan for the whole earth. It is therefore vital to have data on a worldwide scale. Scholars, however, must never lose sight of the peculiarities of given parts of the world and the inherent problems in universal comparability.

CHAPTER TWO

PHYSICAL CONSTRAINTS ON THAI AGRICULTURE

Relief, soil and climate are physical constraints which have a major influence on agricultural activities everywhere in the world.

Thailand is fortunate in having extensive areas of favourable relief. Especially is this the case in the Central Plain, an area of great fertility and productivity and the historic and cultural core of Thailand. The North Eastern or Khorat Plateau, on the other hand, is an extensive level area of not excessive elevation, the agricultural potential of which has not been fully utilized for various reasons, some human, some physical. Its level terrain does, however, render it suitable for the large-scale irrigation works that are intended to transform this backward region. Northern Thailand is essentially a mountainous region. The upland areas are the domain of a wide variety of Hill Tribes, practising shifting cultivation, each with their own altitudinal niche. The northern valleys are made productive at the hands of the settled Northern Thai. Southern or peninsular Thailand has a mountainous backbone with relatively short river valleys, and although extensive agricultural land is limited, agricultural production is diverse and greatly supplemented by sea fishing.

A great deal of work remains to be done on the classification of Thai soils. DUDAL and MOORMANN identified and defined the great soil groups of the Central Valley, Khorat Plateau and Chiang Mai Basin (which was taken to be representative of the North as a whole) [1]. Alluvial soils, low humic gley soils, grey podzolic soils and red-yellow podzolic soils occur in all three regions. In addition, the North has reddish-brown laterite soils and reddish-brown latosols. More varied is the pedology of the Central Plain with its grumusols, rendzinas, red-brown earths, reddish-brown laterite soils, non-calcic brown soils and brown forest soils.

KAWAGUCHI and KYUMA did a wide-ranging survey of the fertility characteristics of Thai paddy soils [2]. The soils of the Central Plain and northern valleys are generally very similar and more or less illitic in their clay-mineral composition, while the soils of the North East have a low illite content and some are rather kaolinitic. The organic carbon content of the surface plough layer of the North Eastern paddy soils is generally low. Soils of the North East are almost all deficient in phosphorus; exchangeable potassium, calcium and magnesium, and readily soluble silica are also deficient. In the North and the Central Plain these minerals are not generally deficient. Salinity is also a problem in the North East.

The soils on which shifting cultivation is carried out are red-yellow podzolic soils, reddish-brown laterite soils, red-brown earths, latosols and alluvial soils [3].

The humus content is outstandingly low on land given over to the cultivation of kenaf, sugar cane, cotton and cassava, while the sandy soils of the natural forest have a humus content as low as 1·64%. On the other hand, the humus content of maize fields is relatively high and rubber fields have a comparatively high and stable humus content. It seems that rubber smallholdings in peninsular Thailand are of value in maintaining soil fertility [4].

There is evidence that current monoculture practices are leading to soil exhaustion particularly in the North East. For instance, the Ban Tum Soil Survey in Kalasin province, North East Thailand, showed rapid deterioration of soil following only a few years of monoculture with cassava and in some cases extensive and irreperable soil erosion and gulleying [5].

Regional variations in temperature in Thailand are not particularly significant in the context of agricultural regions. The data presented in *Table 1* are representative of the whole Kingdom and indicate very little variation in hot-season temperatures. In the cool season the northern meteorological station of Chiang Mai appears cooler, but even here the difference is slight.

Table 1. Temperatures for seven representative meteorological stations

Station	Region	Mean max. °C	Mean min. °C
Phra Nakhon (13°44′ N 100°30′ E)	Metropolitan	32·1	23·3
Chanthaburi (12°37′ N 102°70′ E)	East	30·7	22·5
Nakhon Sawan (15°48′ N100°10′ E)	Upper Central Plain	33·0	22·0
Chiang Mai (18°47′ N 98°59′ E)	North	31·0	19·0
Udon Thani (17°26′ N 102°46′ E)	North—North East	30·7	21·3
Nakhon Ratchasima (14°58′ N 102°07′ E)	Near North East	32·2	20·9
Songkhla (07°11′ N 100°37′ E)	South	31·4	23·6

(Source: Office of the Prime Minister, Meteorological Dept., official records, 1971.)

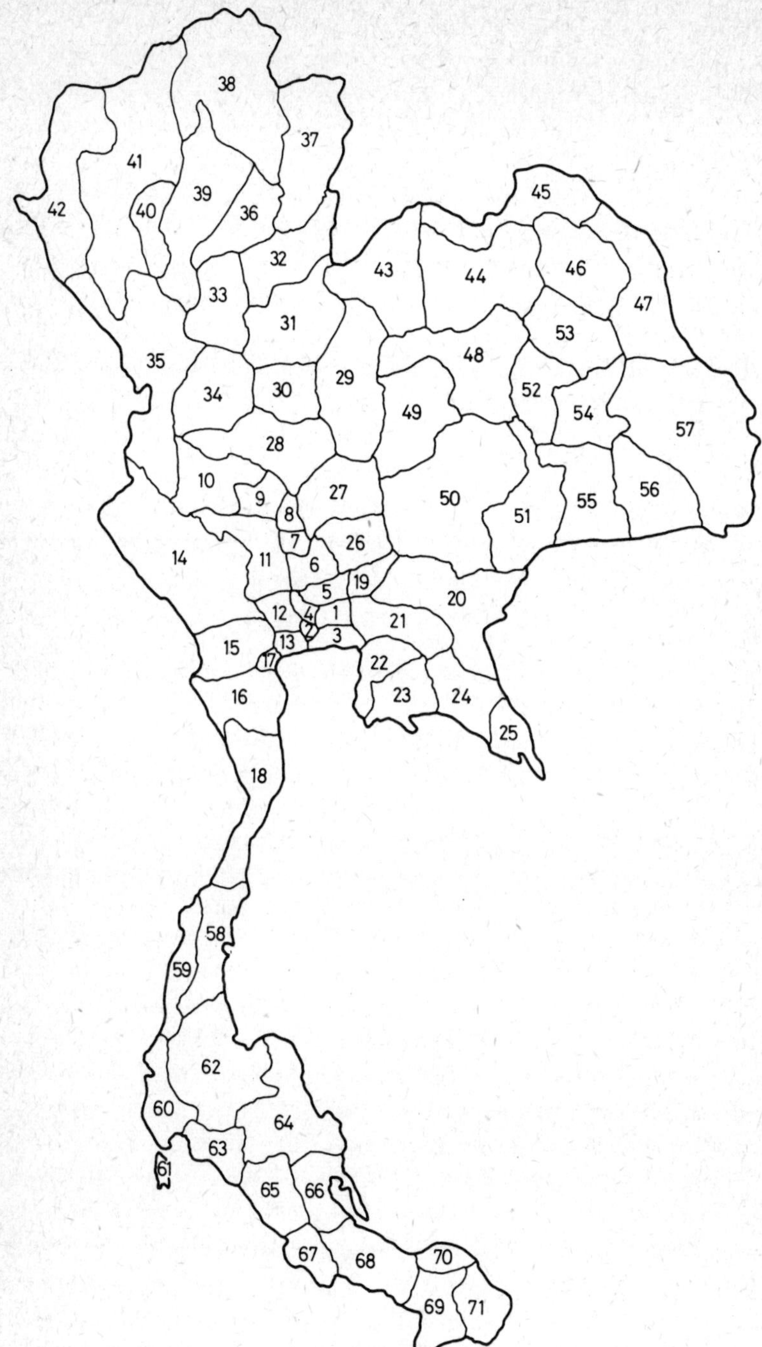

Map 1. Provinces of Thailand

Variation in the intensity and duration of rainfall, on the other hand is more marked, and data for the same seven stations indicates a high incidence of rainy days in the East and peninsular Thailand and markedly lower incidences on the Upper Central Plain and in the North East (see *Table 2*).

Table 2. Number of rainy days at seven representative stations

Station	Number of rainy days
Phra Nakhon	132
Chanthaburi	167
Nakhon Sawan	104
Chiang Mai	127
Udon Thani	119
Nakhon Ratchasima	109
Songkhla	148

(Source: Office of the Prime Minister, Meteorological Dept., official records, 1971.)

Map 2, an isohyet map of Thailand, shows an interesting pattern which has a bearing on subsequent agricultural maps. Two rainfall peaks are evident. One is in South East Thailand on the border of Cambodia, peaking in Trat province; the other is in middle-peninsular Thailand, bordering Burma. A marked peak also occurs in extreme Southern Thailand, bordering Malaysia. Lesser peaks occur in the relatively small areas of the mountainous North, bordering Burma, and along the Mekong River, which forms the international boundary between Thailand and Laos. The remainder of the Kingdom is relatively homogeneous in its receipt of rainfall. It might seem surprising that the North East of Thailand, an area of backward agriculture aggravated by water deficits, does in fact receive the same amount of rainfall annually as the highly productive Central Plain. Other factors, however, are at play. First, year by year fluctuations in the North East are more marked than in the Central Plain. Secondly, the critical date of the onset of the rains and their duration are similarly more variable in the North East. Thirdly, the nature of the terrain and the porosity of the soil permit rapid run-off and loss of water in the North East. Finally, the North East does not possess the impressive irrigation and water-regulation system of the Central Plain.

The humid tropical environment provides ideal conditions for the appearance and diffusion of a wide variety of diseases that affect crops and farm animals, as well as supporting large and injurious populations of insect and animal pests. These indeed are powerful physical constraints on Thai agriculture.

16

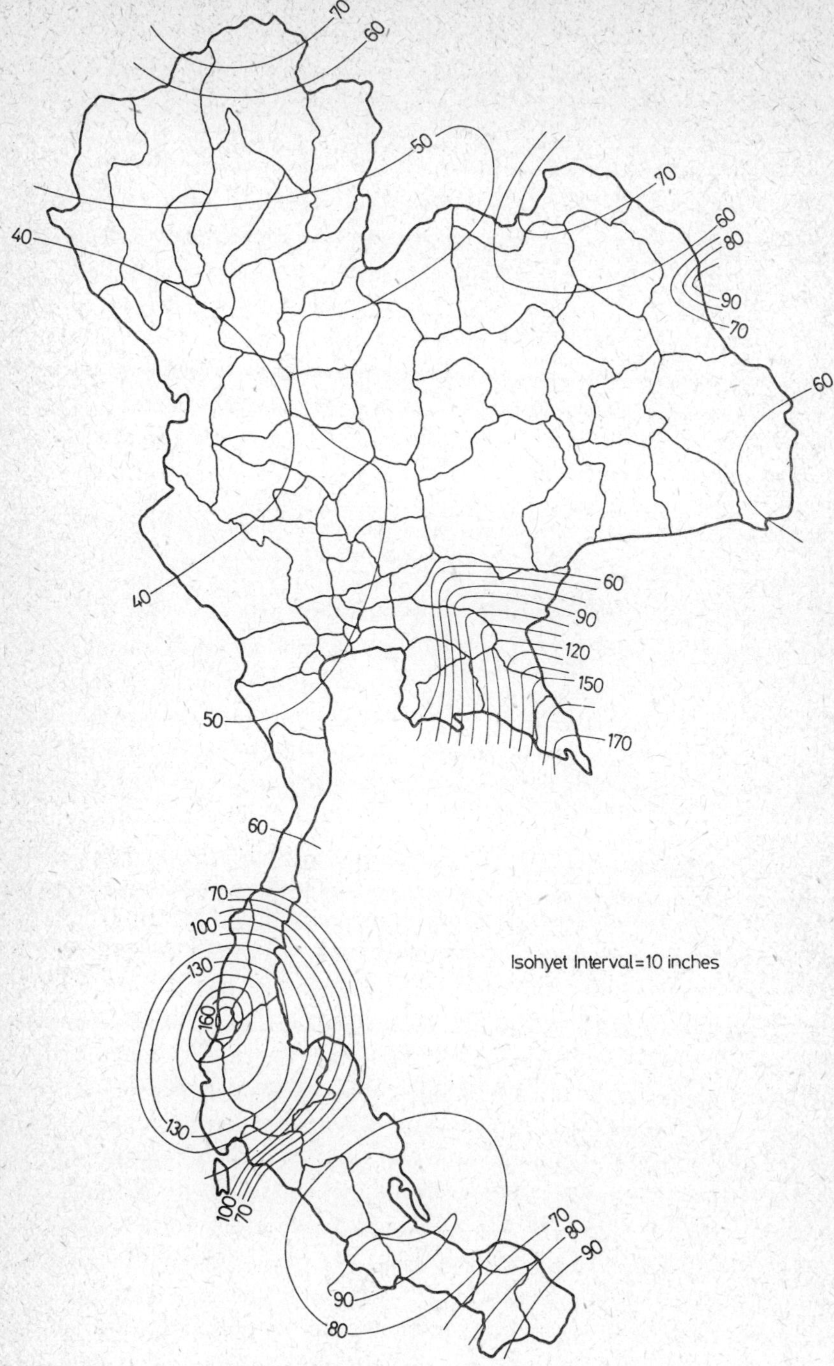

Map 2. Thailand's mean annual rainfall, 1968–71
(Source: Compiled from Meteorological Dept. data)

Insect and animal pests, diseases and viruses are responsible for considerable crop damage, especially that of rice. The principal insect pests are the seedling cutworm, *Spodoptera mauritia* (a leaf eater), the white borer, *Tryporyza incertulas* and *Sesamia inferens* (a stem borer) [6]. There are two serious animal pests, crabs and rats, which take a heavy toll of the rice harvest, as does the rice sparrow, all three of which are hunted by the farmer (or more usually his children) and may supplement the family diet, especially in the poor North East.

The bacterial diseases of rice are as follows: *Piricularia oryzae* (rice blast), *Sclerotum oryzae* (stem rot), *Helminthosporium oryzae* (brown leaf spot), oriental sheaf and leaf spot, collar rot, bacterial leaf stripe, bacterial leaf blight, *Enyloma oryzae* (leaf smut), *Neovessia horrida* (kernel smut), false smut, root knot nematode and rice stem nematode [7].

There are four principal virus diseases of rice, which are yellow-orange leaf virus disease, orange leaf, yellow dwarf and grassy stunt [8].

All field crops suffer from pests and diseases to some extent, but cotton probably of all. For certain varieties that lack resistance to the leaf-sucking jassid *(Empoasce devastans)* cultivation is virtually impossible without a complete spray programme. Maize is affected from time to time by locusts and grasshoppers, pests which were not reported in Thailand until the mid-1950s [9].

Animals are also plagued by disease. Anthrax, haemorriac septocaemia, foot and mouth disease and rinderpest affect cattle and water buffaloes. Pigs are affected by swine fever and swine plague, while poultry are affected by Newcastle disease, smallpox, cholera and infected trachea.

The physical constraints on agriculture can be overcome to a certain extent. Poor soils may be improved by chemical fertilizer, but Thailand is at present one of the lowest fertilizer users in Monsoon Asia, and costs, as was demonstrated by SANSOM, are correspondingly high [10]. Animal manure and night soil have never been prominent, in contrast with their widespread use by Chinese farmers both in China and in overseas communities. Although the use of insecticides and pesticides is still in its infancy, the vaccination of livestock is making good progress and in 1968 90% of all water buffaloes, 93% of all cattle and 95% of all pigs in the North East were vaccinated against rinderpest [11].

Nevertheless, water supply remains the most serious of all the problems faced by Thai agriculture and its control remains potentially the most rewarding way of combatting Nature's disadvantages. This is demonstrated by the fact that in 1968 out of the total area of 2,202,271 acres of rice damaged 4·8% was caused by floods, 83·7% by drought, 8·17% by disease and insects, 1·8% by animals and 2·13% other sources [12].

Irrigation has been the response to water shortage and Northern Thailand has a seven-hundred-year-old tradition of irrigation, whilst the Central Plain had a canal system by 1850 [13]. However, it is the North East which suffers most regularly and severely from rainfall fluctuations and here irrigation is not a significant factor in production. Accordingly, a Co-ordination Committee composed of representatives of the four riparian countries (Thailand, Laos, the then South Vietnam and the Khmer Republic) was set up in 1957 to supervise the integrated development of the Lower Mekong River Valley. This ambitious scheme involved a plethora of international organizations, including SEATO (the South East Asia Treaty Organization), ECAFE (the Economic Commission for Asia and the Far East), FAO (the Food and Agricultural Organization of the U.N.), the World Bank, UNESCO (the U.N. Educational, and Cultural Organization), ILO (the International Labour Organization) and WHO (the World Health Organizaton). Its scope was far-ranging and its projections ambitious as *Table 3* demonstrates.

Table 3. Increase in irrigated land area (double crops) required to meet anticipated demand for agricultural produce 1971–2000 cumulative to end of each year (Units: thousand hectares)

Year ending	Khmer Republic	Laos	N.E. Thailand	Republic of Vietnam
1965 (base year)	0	0	0	0
1970	15·5	5·0	43·0	16·0
1975	35·0	12·5	92·5	34·0
1980	78·0	20·0	187·0	81·5
1985	132·0	28·5	283·0	130·0
1990	208·5	38·5	420·5	206·5
1995	324·0	51·5	600·0	322·5
2000	508·5	84·5	877·0	477·5

(Source: Mekong Secretariat, Bangkok, 1968.)

It was envisaged that this irrigated land would produce a wide variety of cash crops, such as rice, maize, groundnut, soya beans, mung beans, sesame, fruit, vegetables, potatoes and sugar cane. Although North East Thailand was planned as the region in which most development would take place, the vagaries of war and political instability in Indo-China has meant that the Mekong Secretariat has been compelled to concentrate all its developmental efforts there. Nevertheless, in 1971 the North East still had only 1·4% of the total irrigated area in the Kingdom. A thorough evaluation study of the Mekong River Project remains to be done, but small-scale studies indicate the difficulties inherent in the process of modernization and innovation consequent upon irrigation in North East Thailand [15].

REFERENCES

[1] Dudal, R. and Moormann, F. R., "Major Soils of South East Asia," *Journ. Trop. Geography*, Vol. 18, pp. 54–80, 1964.

[2] Kawaguchi, K. and Kyuma, K., *Lowland Rice Soils in Thailand*, Centre for South East Asian Studies, Kyoto, 1969.

[3] *Report of the United Nations Survey Team on the Economic and Social Needs of the Opium Producing Areas in Thailand*, 1967.

[4] Sato, T., *Field Crops in Thailand*, Centre for South East Asian Studies, Kyoto, 1966.

[5] Bradnock, R. W., O'Reilly, F. D. and Stott, P. A., *Ban Tum Soil Survey*, School of Oriental & African Studies, London, 1972 (unpublished).

[6] Chakratong Tongyai, M. R., *Review of Pests of Rice in the Field in Thailand*, Min. of Agr., Bangkok, Mimeo.

[7] Worawisithumrong, A. and Sithchai, T., *Rice Diseases in Thailand and Control Measures*, Technical Division, Rice Dept., Min. of Agr., Bangkok, Mimeo.

[8] Wathanakul, L. *Rice Virus Diseases in Thailand*, Rice Protection Research Centre, Technical Division, Rice Dept., Min. of Agr., Bangkok, Mimeo.

[9] *The Agricultural Economy of Thailand*, U.S. Dept. of Agriculture, Economic Research Service, 1972.

[10] Sansom, R. L., *The Economics of Insurgency in the Mekong Delta of Vietnam*, MIT Press, 1970.

[11] *Annual Report of Livestock*, Livestock Dept., Min. of Agr., Bangkok, (in Thai).

[12] *Annual Report on Rice Production in Thailand*, Min. of Agr., Bangkok.

[13] Tanabe, S., *Historical Development of the Canal System in the Chao Phraya Delta*, Part 1, Kyoto Univ., Japan 1973 (in Japanese).

[14] Royal Irrigation Department Handbook, Bangkok.

[15] O'Reilly, F. D., *The Process of Innovation in the Subsistance Agriculture of North East Thailand with Particular Reference to the Lam Pao Irrigation Area, Changwat Kalasin*, Ph. D. thesis, School of Oriental and African Studies, Univ. of London, 1974 (unpublished).

CHAPTER THREE

POPULATION AND THE LAND

Overpopulation, considered the scourge of Monsoon Asia, is not as yet a major problem in Thailand. In 1970 the population of the whole Kingdom was 34,152,000 on an area about the size of France, giving a population density, according to the writers' estimates, of 66 persons per sq. km. This contrasts sharply with the population densities in 1963 for India (151·2 persons per sq. km.), Japan (159·4), Vietnam (100·5), Indonesia (67·1) and the Philippines (100·8). Clearly, Thailand possesses a favourable overall man–land ratio.

The writers estimate the population density per crop acre in 1970 for the whole Kingdom to have been 1·4 and the population per rice acre 1·9. The latter figure is particularly important, since it expresses the ratio between the population and its staple food resource base, and shows that there is no great pressure on the nation's rice lands, which should therefore be able to sustain adequately its population. Rural population densities were similarly favourable, the farm population per crop acre being 1·1 and per rice acre as low as 0·8 in 1970.

INGRAM stated that in 1950 only 11% of the total area of the country was cultivated [2], rising, according to the writers' estimate to 19% in 1970, while 60% of the nation's population was in agricultural households. Government sources state that 80% of the Thai population is engaged in agriculture, forestry and fishing. The discrepancy between the two figures results not only from the government's broader classification, but also from the fact that it is concerned with the labour forces, whereas the writers are dealing with categories of household. Official sources state that 70% of the country is still forested.

It is clear, therefore, that Thailand is not overpopulated: the farming population enjoys a favourable man–land ratio and there is ample space for future agricultural expansion. However, there may be a conflict of interests between the government's attitude to forest as a valuable national asset and the farmers' desire to exploit the forest for firewood, charcoal, construction and agricultural expansion.

The picture nationwide varies somewhat as indicated in *Table 4*. The highest pressure on crop land is in the Upper Central Plain, North and near North East. The lowest pressure on crop land is in the South, East and West. The table also gives the population per paddy acre and per rice acre, the latter figure including upland rice as well as wet rice. The two figures do not differ in the Metropolitan area, Central Plain, West and East, since in these areas upland rice is insignificant. The highest pressure on rice land is found in the North and Near North East; the lowest pressure is in the Central Plain.

Table 4. Population pressure by Thai regions

Region	Mean agricultural population	Mean persons per crop acre	Mean persons per paddy acre	Mean persons per rice acre
Metropolitan	86,338	1·1	1·1	1·1
Central Plain	175,869	1·0	0·7	0·7
West	240,597	0·6	1·0	1·0
East	139,287	0·5	1·4	1·4
Upper Central Plain	272,741	1·6	1·2	1·1
North	367,093	1·6	2·5	2·3
North North East	463,797	1·1	1·7	2·2
Near North East	510,133	1·5	2·4	2·4
East North East	668,237	1·0	1·3	1·4
South	190,973	0·6	2·3	2·1

The figures for the provinces were subdivided into 3 categories: 0·0 to 1·0 persons per crop acre, 1·0 to 2·0, and over 2·0. The majority of all provinces have a density of less than 1 person per crop acre, the island of Phuket having the very low density of 0·03. Twenty provinces fall into the category of 1·0 to 2·0. These are Nonthaburi and Phra Nakhon (Metropolitan), Tak and Uttaradit (Upper Central Plain), Chumphon and Ranong (South), Kalasin, Si Sa Ket and Surin (East North East), Buri Ram, Chaiyaphum, Khon Kaen and Nakhon Ratchasima (Near North East), all the provinces in the North North East and all the provinces in the North except Lampang. Three provinces have densities of over 2 persons per crop acre: Lampang (North), Phetchabun (Upper Central Plain) and Samut Songkhram (West). The last two have abnormally high densities, Phetchabun's figure being 9·1 and Samut Songkhram's 8·7.

The provincial figures for population density per rice acre were subdivided into five categories: 0·0 to 1·0 persons per rice acre, 1·0 to 2·0, 2·0 to 3·0, 3·0 to 4·0, over 4·0. Table 5 shows the results. The vast majority of provinces have a density of below 3 persons per rice acre and there is a fairly even spread in the first three categories. Phuket's density of 3·6 persons per rice acre contrasts with that previously cited, of 0·03 persons per crop acre, indicating extensive field crop cultivation. Similarly, in Yala, in the far South, the rice acre density of 3·5 contrasts with the crop acre density of 0·3 for the same reason. In the category above 4·0 are three provinces with extensive non-rice areas, resulting in relative demographic pressure on their rice areas. Samut Songkhram in the West has the abnormally high density of 15·1 persons per rice acre.

Table 5. Population density per rice acre

Category	Provinces
0·0 to 1·0 persons per rice acre	Ang Thong, Chachoengsao, Chai Nat, Kamphaeng Phet, Nakhon Nayok, Nakhon Sawan, Pathum Thani, Phichit, Phitsanulok, Phra Nakhon Si Ayutthaya, Roi Et, Sakon Nakhon, Samut Prakan, Samut Sakhon, Sara Buri, Sing Buri, Suphan Buri, Thon Buri, Uthai Thani,
1·0 to 2·0 persons per rice acre	Chiang Rai, Chon Buri, Kalasin, Kanchanaburi, Lop Buri, Maha Sarakham, Nakhon Pathom, Nakhon Phanom, Nong Khai, Nonthaburi, Pattani, Phangnga, Phatthalung, Phetchabun, Phetchaburi, Phra Nakhon, Ratchaburi, Satun, Si Sa Ket, Songkhla, Sukhothai, Surat Thani, Trat, Udon Thani, Uttaradit.
2·0 to 3·0 persons per rice acre	Buri Ram, Chaiyaphum, Chanthaburi, Chiang Mai, Chumphon, Khon Kaen, Krabi, Lampang, Lamphun, Mae Hong Son, Nakhon Ratchasima, Nakhon Si Thammarat, Nan, Narathiwat, Phrae, Ranong, Rayong, Surin, Tak, Trang, Ubon Ratchathani,
3·0 to 4·0 persons per rice acre	Phuket, Yala
Over 4·0 persons per rice acre	Loei, Prachuap Khiri Khan, Samut Songkhram

In a country with a generally favourable man–land ratio one would not expect to meet food shortages, malnutrition and hunger. Yet OSHIMA reported that the Interdepartmental Committee on Nutrition for National Development (organized by the National Institute of Health of the U.S. Government) did not show any Asian calorie-deficiencies except for Thailand. He further considered that a low calorie intake in Thailand is associated with a low multiple cropping index, since insufficiency of calories expresses itself in inadequate work effort after the peak season [3].

O'REILLY certainly found food shortages in Kalasin province. In an area that he estimated as 67% fully subsistent, he found that 8% of the farms failed to satisfy the modest subsistence requirements of 160 kg per capita per annum. Physical manifestations of malnutrition were observed, especially in young children [4]. It is also likely that Hill Tribe populations receive an inadequate diet, aggravated by excessive opium indulgence. Of course, there are some provinces of the Kingdom as in the Upper Central Plain and the West, where farmers are so cash-crop orientated that they fail to produce sufficient rice; but this is a fundamentally different situation from that in the North East, since these favoured cash crop farmers are assured of a regular annual income from cash-crop sale with which to purchase rice on the open market for familial consumption, whereas the poor North Eastern farmer, having neither rice nor cash, must perforce borrow either rice or cash in order to survive.

Although Westerners carp at the Thai preference for polished rice instead of the more nutritious practices of consuming the whole grain or parboiling the rice, it seems nevertheless that Thai dietary preferences, emphasizing as they do fish, fruit and vegetables, provide the wherewithal for a balanced healthy diet, provided the commodities are perennially and sufficiently available. The ancient Thai maxim, taken from a stone inscription of the 13th century in Sukhothai that "In the water there are fishes, in the paddy fields there is rice," sums up the Thai ideal of peace and prosperity.

Agricultural households are not only consumption units, however, but are also the sources of agricultural labour. In Thailand both sexes and all ages are engaged in agriculture. In some tribal communities agriculture is entirely the domain of women. The question arises as to the degree of unemployment or underemployment in Thai agriculture. MOTOOKA is convinced that severe unemployment exists in Thai agriculture. He computes a national average of 100 working days per head of population engaged in paddy cultivation per annum, divided among nursery care and paddy preparation from March to May, transplanting from May to June, harvesting from October to December, and "slow threshing" from December to March. He calculates second crop paddy as only 1% of the total paddy land and considers that other crop and livestock enterprises make few demands upon labour [5].

O'REILLY, however, found that in Kalasin province higher labour inputs were associated with greater yields on both rice and kenaf plots, indicating that the marginal productivity of labour there has not yet attained zero. However, there is undoubtedly seasonal unemployment, and to a greater or lesser degree labour is concentrated in the months from April to July and from October to November [6].

If there is rural unemployment during the dry season then there is clearly scope for expansion of irrigated agriculture during that season. There is evidence, however, that the dry season is not an idle period, since farmers or their children travel to seek off-farm work in the towns and may, as in construction work (undertaken by both men and women), work harder and more regularly than they do on the farm. Dry-season irrigated farming must be more profitable to the farmer than off-farm urban work before it can win acceptance and play its essential role in diversifying the agricultural base of the Kingdom.

INGRAM has suggested that the rate of population increase has been roughly the same in all parts of the country and that regional shifts in population do not appear to have been very significant [7]. However, migratory harvest labour is a feature of the rural scene, both traditional and contemporary. Thus North Eastern farmers travel from their home villages every year to specific Central Plain villages to work as agricultural labourers on the rice harvest. At present, there is evidence of more permanent rural-to-rural migration, notably from the North East to the more prosperous Upper Central Plain. This phenomenon, however, awaits an authoritative analysis.

At the same time, urbanization is preceding apace in Thailand, especially the growth of the capital, Bangkok which, with its 3 million or so people, provides a striking example of urban primacy. The GOLDSTEINS say that from 1947 to 1970 Thailand's urban population increased from 10% to 13% of the total and that the number of urban places containing 20,000 or more persons grew from 6 to 38. The rate of urban growth has been high, averaging about 5% per year, which is just above the average for the world's developing countries, but since its rural growth rate has also been quite high, about 3%, the rapidity of the urbanization process does not appear as marked as elsewhere in the Third World [8].

From an analysis of the 1960 and 1970 Population Censuses O'REILLY found that migration from the provinces to Bangkok was to a great extent a distance-decay phenomenon, the number of migrants per 10,000 persons in the home province declining with distance from the capital. At the same time there appears to have been a recent increase in outmigration from the economically poor provinces of the North East and from strategic border provinces of the North and South, plagued by insurgency [9].

The impact of rural to urban migration on the capital is immediately apparent and has prompted a great deal of writing, both academic and popular. However, detailed analytical studies of the effect of rural to urban migration on agricultural production and on the rural social structures of those provinces which are losing population remain to be made.

REFERENCES

[1] Motooka, T., *Agricultural Development in Thailand*, Discussion Paper No. 27, Centre for Southeast Asian Studies, Kyoto University, Kyoto, Japan.

[2] Ingram, J. C., *Economic Change in Thailand 1850–1970*, Stanford Univ. Press, 1971.

[3] Oshima, H. T., "Food Consumption, Nutrition and Economic Development in Asian Countries", *Economic Development and Cultural Change*, Vol. 15, No. 4, July 1967.

[4] O'Reilly, F. D., *The Process of Innovation in the Subsistence Agriculture of North-East Thailand with Particular Reference to the Lam Pao Irrigation Area, Changwat Kalasin*, Ph.D. thesis, S.O.A.S., Univ. of London, 1974 (unpublished).

[5] Motooka, *op. cit.*

[6] O'Reilly, *op. cit.*

[7] Ingram, *op. cit.*

[8] Goldstein, S. and A., "Types of Migration in Thailand in Relation to Urban-Rural Residence", Working Paper, Population Studies and Training Centre, Brown University, Providence, Rhode Island, U.S.A.

[9] O'Reilly, "Migration to Thailand's Primate City 1960–1970", Working Paper, Dept. of Geography, Bayero University, Kano (Nigeria), 1978 (unpublished).

CHAPTER FOUR

THE STRUCTURE OF THAI FARMING

Agriculture takes place within a social as well as physical framework. Within the traditional agriculture of the developing world farming is more than an occupation; rather it is a sub-culture. Land, similarly, is more than an economic commodity to be bought and sold at random; rather it is a specific and precious possession which not only supports but also gives an identity to the rural community, being intricately interwoven with its life through complex socio-economic and ritualistic threads. These considerations may have an important impact upon land-use. In the first place, traditional societies did not normally "mine" their soil or exploit it to exhaustion. Rather they viewed it as a valuable and delicate ecosystem to be passed on intact, enhanced rather than depleted, to prosperity. This attitude, belatedly lauded by Western conservationists, is now fading under the impact of the cash economy. Secondly, the debate continues in all branches of social science as to whether the peasant is "an economic man". Does he seek to maximize profit like the economic man of classical economic theory, or is his mind governed by sub-rational ritualistic and/or social considerations? The writers' observations on the Thai situation are that the prime motivation of the peasant is economic, i.e. the sustaining of himself and his family from the produce of his land. Once subsistence has been achieved, however, a wide variety of conflicting motivations may come into play.

Monsoon Asia was commonly regarded as an area of severe, chronic and iniquitous landlordism. Thailand, however, does not share and has never shared this problem with India. On the contrary, Thailand is primarily a land of owner-occupiers, as *Map 3* shows. The relatively backward North Eastern provinces are *par excellence* the domain of the owner-occupier. Conversely, it is in the productive lower Central Plain, especially in the provinces around Bangkok, that cash renting is significant. Cash renting here is a traditional feature, although there is some evidence that it is on the increase. Noticeable also is the relatively high percentage in the coastal province of Chon Buri, a traditional area of intensive garden cultivation by immigrant Chinese. Falling within the range 5% to 9·9% are Ratchaburi, Phra Nakhon Si Ayutthaya and Chachoengsao, which is comprehensible, since they border the cash-rented core. However, Sukhothai stands out in the northern Central Plain as an isolated, relatively higher area of

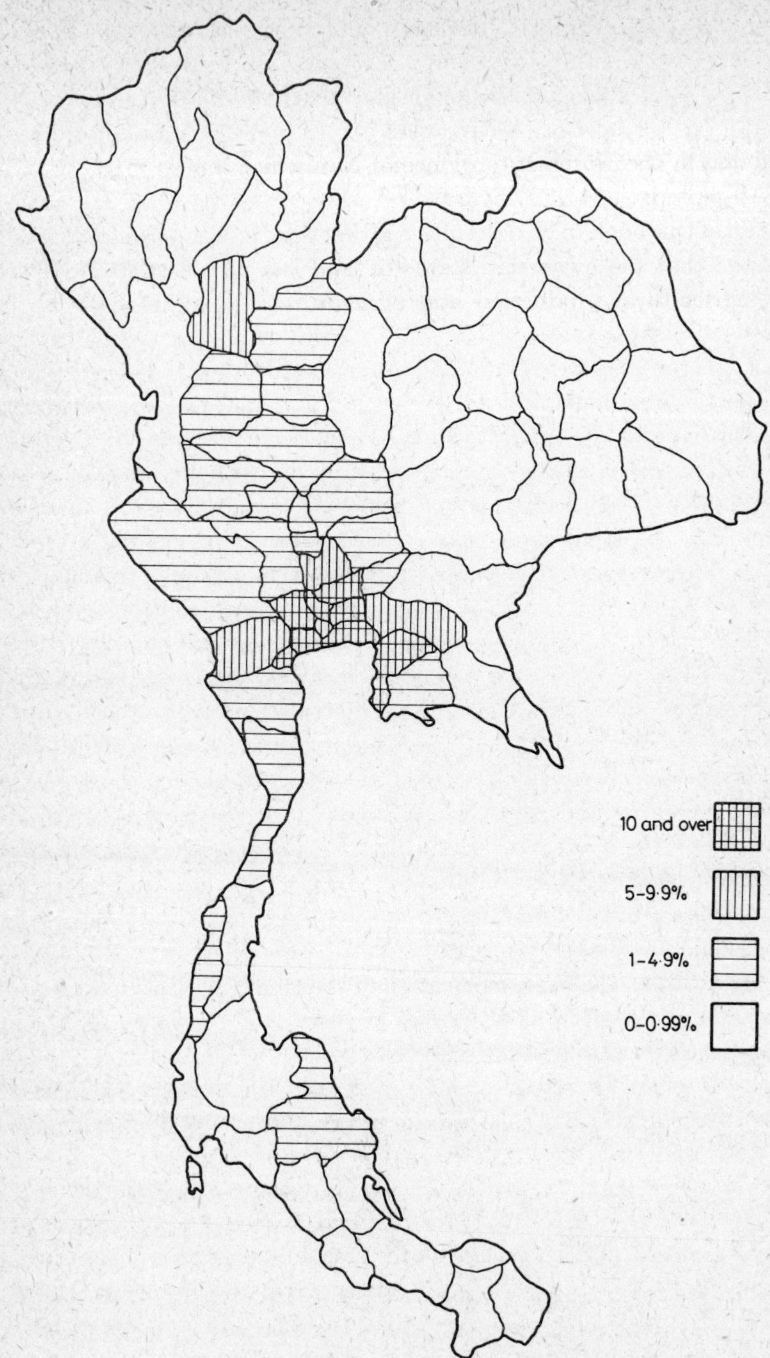

Map 3. Cash rented holdings as percentage of total holdings (Source: Census of Agriculture)

cash renting. Most of the Central Plain and parts of the peninsula fall within the category 1% to 4·9%. A sense of perspective must be maintained, however, because even in the province with the highest incidence of cash-renting, Thon Buri, only 23% of all holdings are cash-rented. The lowest incidence of cash-renting is found in the Northern province of Nan which has only 0·03% of its holdings cash-rented.

It is the aim of this book to be descriptive rather than polemic. Nevertheless, it must be stated that the ownership status of land is a not insignificant factor influencing agricultural production and agricultural development. That the Central Plain is more productive than the North East is not an indication that cash-renting *per se* is more productive than owner-occcupation. Rather it demonstrates the phenomenon that in a commercial and highly productive rural economy the value of land increases. Given the free play of market forces, some farmers will achieve greater prosperity than others, whilst others will decline in wealth absolutely as well as relatively. The scene is then set for the wealthier farmers to expand their agricultural holdings at the expense of the poorer ones, thus reducing the latter to landlessness. Urban businessmen may also enter the land-market for highly profitable farm land, thus giving rise to the absentee landlord. Which system, owner-occupation or cash-renting, is more conducive to agricultural development? The owner-occupier is psychologically more amenable to the improvement of his land holding than the renter, who perhaps has little security of tenure, and is essentially improving for another man. On the other hand, the owner-occupier may not have the capital and the technical expertise to effect improvements, commodities in greater supply among the landlords. Similarly, the small owner-occupier, living from hand to mouth, may not dare to risk change or innovation; instead, he sticks doggedly to the old tried system which assures reliable if small returns. On the other hand, the landlord, perhaps with a number of holdings, has sufficient security to risk with impunity innovation and charge. The basic motivation of the landlord is maximization of cash income; the motivation of the small peasant owner-occupier is feeding himself. Reality may make these arguments somewhat academic, since it seems that in markedly capitalist societies (such as Thailand) increasing landlordism and landlessness are a necessary concomitant of agricultural development, unless specific government intervention prevents this inexorable process.

The size of holding is a significant factor influencing agricultural production and development. *Map 4* shows the distribution of large holdings in Thailand, large holdings in this context (the writers' delimination) being those between 24 acres and 55 acres. The core area of large holdings (with over 15% of all holdings large) comprises the adjacent provinces of Pathum Thani, Nakhon Nayok and Chachoengsao. Provinces in the next category (10%–14·9%) fringe this core to

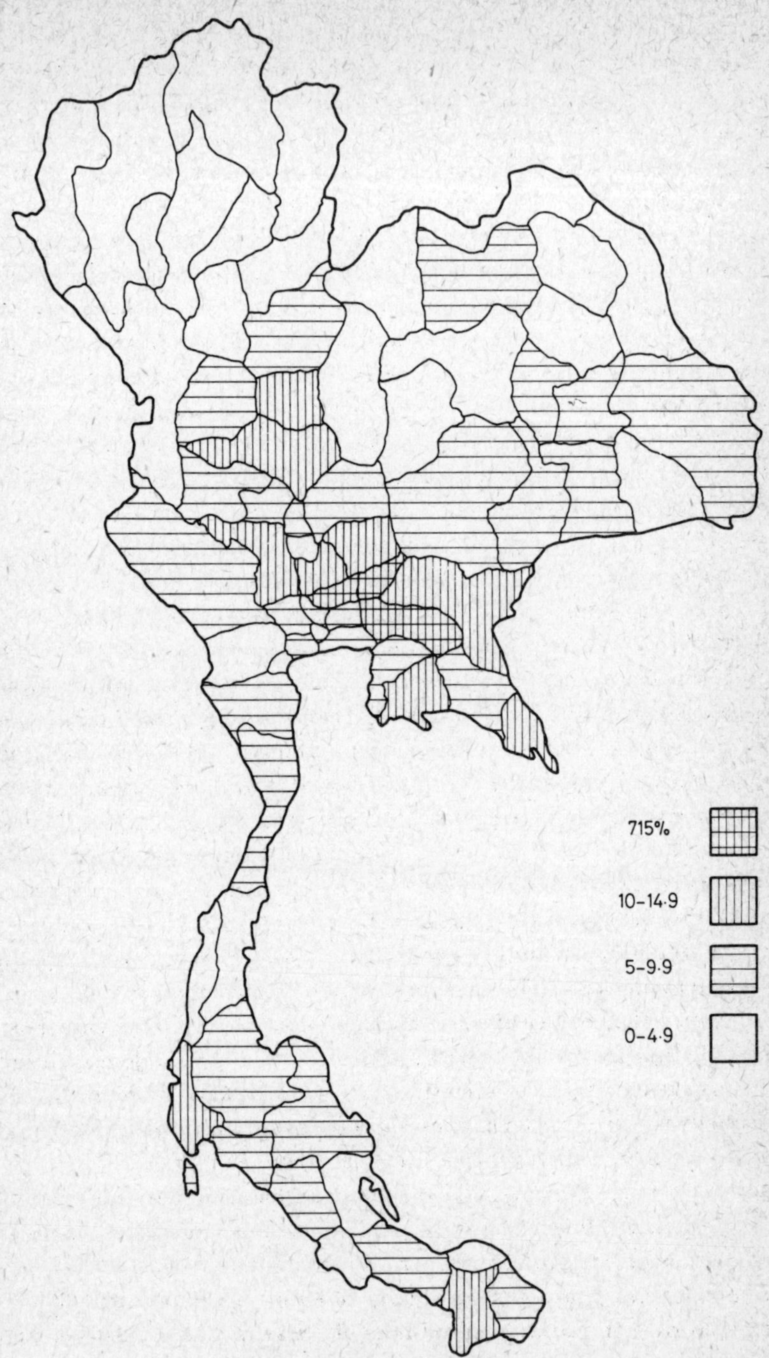

Map 4. Large holdings as percentage of total holdings (Source: Census of Agriculture)

the North but also occur in the East in Rayong and Trat, in the peninsula in Phangnga and Yala and in the Upper Central Plain in Nakhon Sawan and Phichit. The lowest proportion of large holdings is found in the Northern provinces, in a substantial group of North Eastern provinces and in some peninsular provinces. Metropolitan Thon Buri and, anomalously, Ang Thong, fall into this category.

Map 5 is in some ways the opposite of the preceding map, and shows the distribution of small holdings in Thailand, i.e. holdings under 6 acres. The core area (with over 75% of all holdings small) is the Northern bloc of provinces. Loei, an outlier belonging geographically to the North East but economically to the North, falls into this category, as does the tiny coastal province of Samut Songkhram. No province in the Kingdom has less than 25% of its total holdings below 6 acres in size. The majority of the provinces fall into the category 25% to 50%, while those belonging to the category 50% to 75% constitute a dispersed and heterogeneous group. In the Upper Central Plain are Uttaradit and Phetchabun, in the North East Nong Khai, Chaiyaphum and Nakhon Phanom, and in the peninsula Ranong, the island of Phuket and the rubber-producing provinces of the far South, Songkhla, Pattani and Narathiwat.

On the whole, then, Thailand is primarily a country of small holdings. Pathum Thani is the province with the highest concentration of large holdings, yet only 26% of its holdings are large, and in the same province almost as many holdings (23%) are small. Thus even the big farm areas still have sizeable proportions of small farms. In Nan in the North fully 95% of all holdings are small, while the figures for the Northern provinces of Chiang Mai (90%), Mae Hong Son (92%) and Lampang (91%) are similar.

Clearly, there are spatial inequalities in farm size in the Kingdom as a whole, whose explanation may be partly historical and partly geographical. Thus the large holdings of the Central Plain are a result of the highly commercial and highly productive natures of the agriculture of that area and relative importance of landlordism. The small holdings of the North are a response to population pressure on the restricted agricultural land of the valley floors, hemmed in by the northern mountains. Inequality in farm size may also be found at the village level in all regions, as O'REILLY has shown in a study of seven North Eastern villages in Kalasin province [1].

What effect does size of holding have upon agricultural production and development? It is commonly believed that small holdings are farmed more intensively than larger ones and thus have a higher yield per acre. On the other hand, larger holdings have a larger total product. Being more profitable and having more capital, the larger farmer can utilize technical inputs which give him a higher yield per manday than his smaller counterpart relying upon family

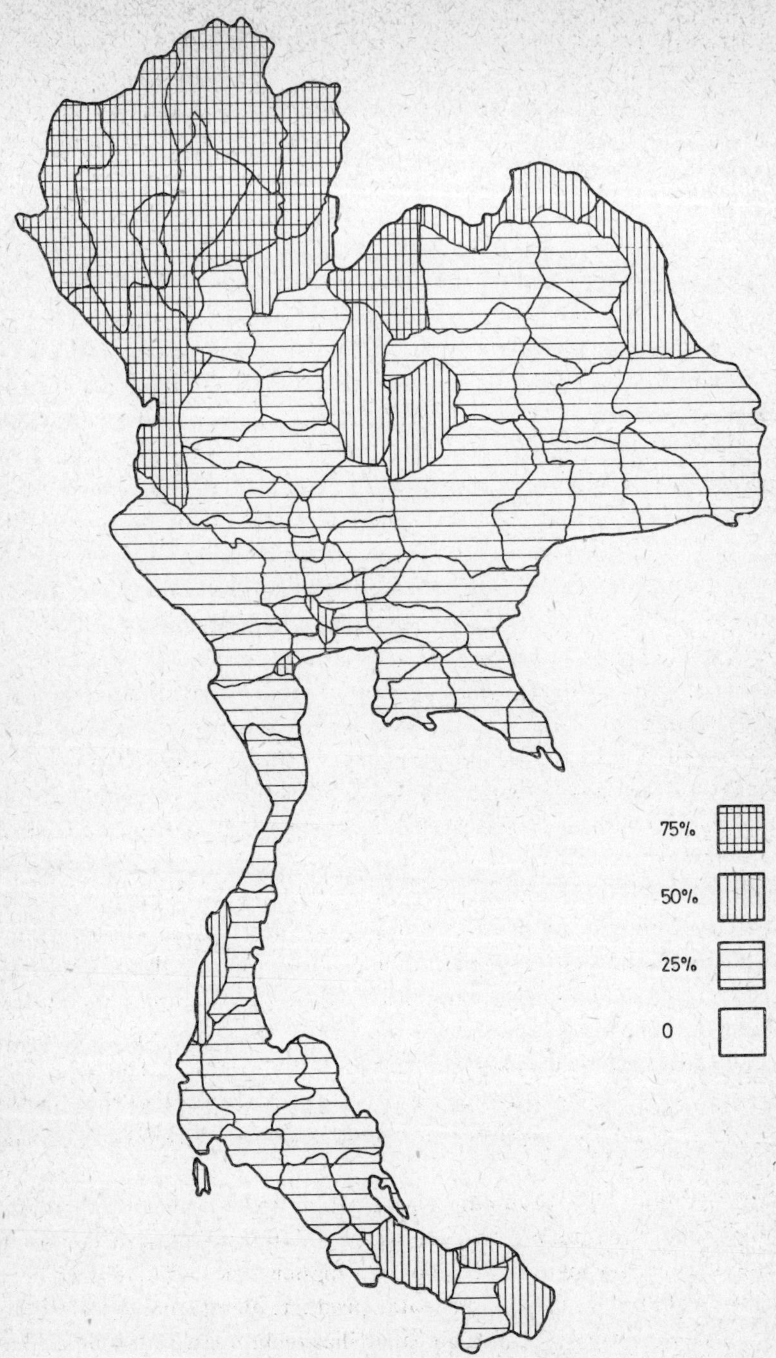

Map 5. Small holdings as percentage of total holdings (Source: Census of Agriculture)

labour. Moreover, mechanization and the use of expensive chemical inputs make more sense on the large holding.

It may be suggested that the emergence of the large holding and the closely related emergence of landlordism are the necessary concomitants of the process of modernization in capitalist developing countries, unless specific government legislation prevent them. Although the end result may be a more modern and more productive agricultural sector, the negative results may be the emergence of a rural landless labouring class or, worse still, an unemployed community of aimless and unskilled people. Such people, alienated from their land, may easily become alienated from political authority and constitute a potentially dangerous, disaffected populace.

Mechanization and the use of chemical fertilizers are often regarded as indicators of modern commercial agriculture and *Map 6* shows the distribution of holdings using chemical fertilizers as a percentage of total holdings. A core area (with over 20% of holdings using chemical fertilizers) emerges around the metropolitan area. Kamphaeng Phet is an interesting outlier of the core area. The second group (10%–19·9%) is found in certain provinces adjacent to the core, such as Nakhon Pathom, Samut Songkhram, Ratchaburi and Chanthaburi, but also in Phichit (adjacent to Kamphaeng Phet) and further north in Lamphun. The third category is scattered through the Central Plain, North East and peninsular. The North East presents a solid bloc as a fertilizer deficit area.

Map 7 shows the distribution of holdings using tractors as a percentage of total holdings. The metropolitan area does not stand out here, probably because the holdings are not large enough and a group of provinces in the Central Plain and East form something of a core with Rayong, Phichit and Kanchanaburi as outliers. The second group (10·0–19·9%) constitutes a bloc in the Central Plain and a smaller bloc in the southern peninsula, with Chon Buri in the East and Prachuap Khiri Khan in the West as outliers. Only a few scattered provinces belong to the third category (5·0%–9·9%), the bulk of them falling into the last category (0–4·9%). The North East and most of the North are shown as unmechanized areas.

Thus both chemical-fertilizer and tractor use are still relatively unimportant in Thailand. The use of chemical fertilizers was zero up to World War I, whilst the use of animal manure and night soil has never been significant. In 1963 only 9% of the farmers in the Kingdom applied any chemical fertilizer, whilst an additional 35% used only organic fertilizer [2]. In 1965 60% to 70% of all fertilizer was used on rice and most of the rest on sugar cane, rubber, tobacco, fruits, vegetables and maize [3]. Thus despite a rapid increase since World War I, Thailand still remains one of the lowest users of fertilizer in South East Asia. In

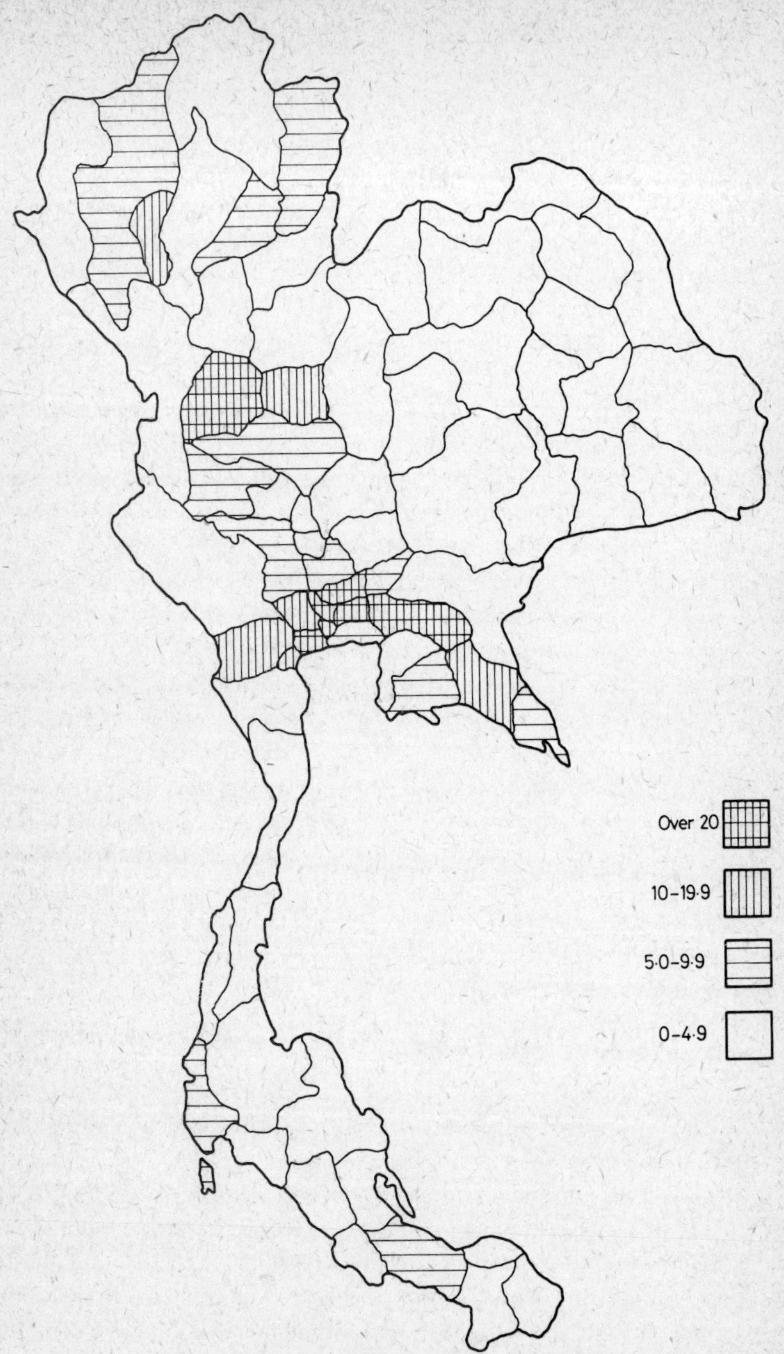

Map 6. Holdings using chemical fertilizers as percentage of total holdings (Source: Census of Agriculture)

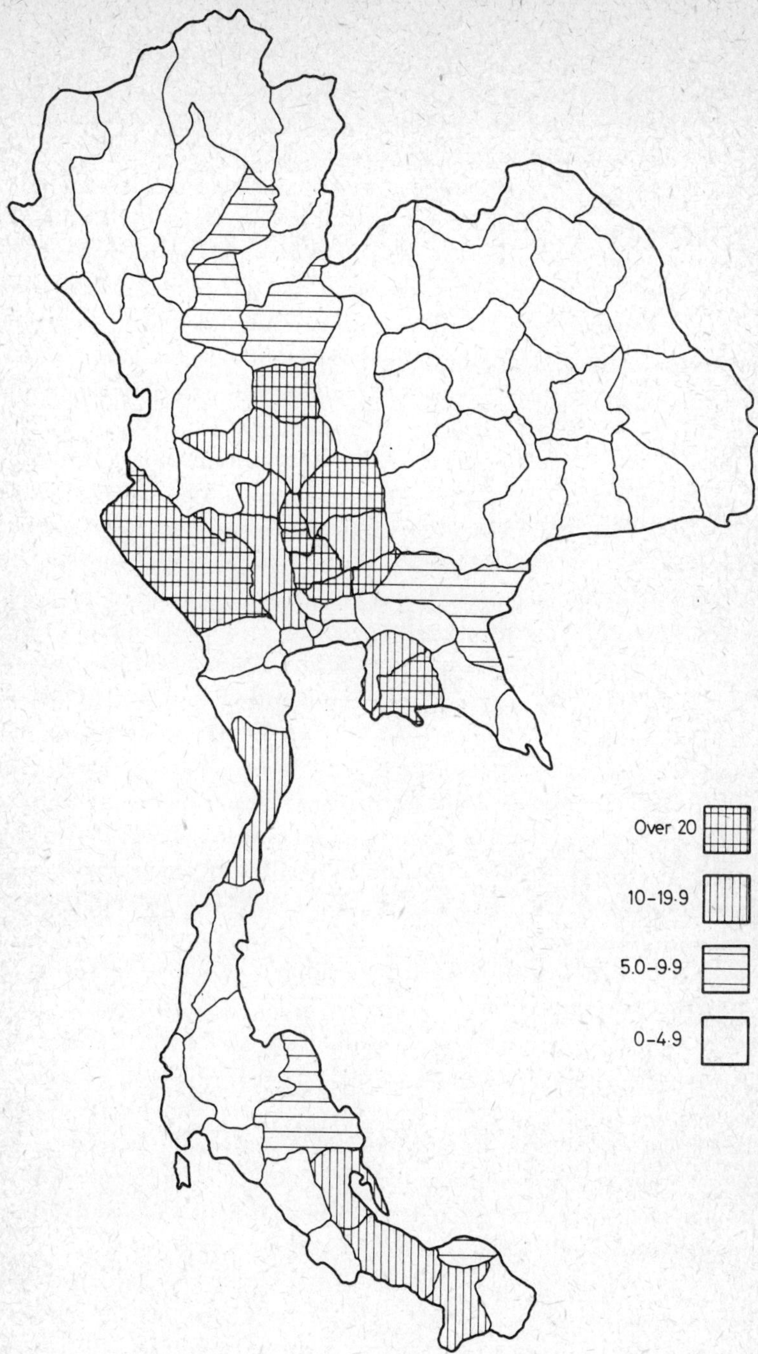

Map 7. Holdings using tractors as percentage of total holdings (Source: Census of Agriculture)

1969 the total nutrient requirements (NPK) were estimated at 272,050 metric tons as against an actual estimated consumption of only 108,700 metric tons [4]. This is not surprising, since fertilizers have long been considerably more expensive in Thailand than elsewhere in the region. Thus SANSOM worked out a price ratio on the basis of a kilogram of nitrate against a unit of paddy for selected countries. His ratios of 1 to 16 for the Republic of Vietnam, 1 to 6 for Pakistan and 1 to 4 for Japan contrast with the ratio of 5 to 9 for Thailand [5].

Most chemical fertilizer used in Thailand has to be imported (mainly from Japan and the Federal Republic of Germany) and the Chemical Fertilizer Co. Ltd at Mae Mo in northern Thailand is demonstrably uneconomic. United States concerns are, however, interested in expanding into the fertilizer industry in Thailand [6].

The question arises as to the relative benefits and desirability of fertilizer use in Thailand. From the plethora of results, some contradictory, on chemical fertilizer application in developing countries, it can be stated without reservation that the benefits are neither ubiquitous nor automatic: the nature of the soil, the exact composition of the chemical fertilizer, the time and amount of application, traditional agricultural practices and the development of adequate irrigation are all constraining factors. Doubtless, many farmers have built up a psychological mistrust of chemical fertilizers after disappointing or even harmful results through application without awareness of these constraining factors, and it is hardly surprising that chemical fertilizers have made so little ground in Thailand. Paradoxically, chemical fertilizers are least important in those areas which would most benefit from their application—the poor, deficient soils of the North East. However, the relevant authorities in Thailand are committed to the spread of chemical fertilizers and it is likely that a great expansion of use will take place in the future.

The impact of mechanization on Thai farming is quite recent. In 1963 out of a total of 3,087,141 holdings in the Kingdom only 3·3% used mechanical power exclusively, whilst 11·6% used both mechanical and animal power and 85% depended entirely upon human and animal power [7]. There being no domestic manufacture of tractors, the country is dependent on the import of tractors in "knocked-down" condition for assembly in Thailand. There are, however, significant domestic manufacturers of tractor-drawn dise ploughs and harrows but, despite this, the import of these articles is still important. The major suppliers are the United Kingdom, Japan and Finland.

When first introduced into Thailand, a government tractor service was provided, but was later discontinued for financial reasons. At present, most farmers who use tractors hire them from private entrepreneurs or from farmers' associations and co-operatives. There is some controversy regarding the effects of tractor use on farm operations for while CHANCELLOR found "a definite relationship between tractor use and the initiation of new farm enterprises" [8], a Thai government co-ordinated study concluded that "the availability of equipment for farm tasks has not significantly affected farming cycles or the extent of multiple cropping as yet" [9].

Nevertheless, it is clear that there is vast scope for increased mechanization in Thailand. The co-ordinated study conservatively estimated Thailand's potential farm tractor population as 80,000 to 120,000 units [10], as against only 17,500 units or 15% to 22% of the potential in the year of the study. A 1972 estimate suggested that the market for agricultural machinery in Thailand was of the value of 200 million U.S. dollars per year and could be expected to grow by about 15% to 20% over the subsequent three years [11].

In brief, although it is generally agreed that mechanization makes labour more efficient, it is by no means agreed that mechanization increases the yield per unit of land. It is also debatable whether in peasant conditions of cheap and ample labour mechanization makes either economic or moral sense. It is clear, nevertheless, that the mechanization process will proceed apace in Thai agriculture, fundamentally affecting both agricultural economics and rural society in the Thai countryside.

REFERENCES

[1] O'Reilly, F. D., *Process of Innovation....*, op. cit.

[2] Bond, B. J., Kelso, M. T. and Woodward, O. R., *A Report on the Thailand Fertilizer Situation and Potential*, TVA, Muscle Shoals, Alabama, U.S.A., Mimeo AID 1966.

[3] *FAO/UNDP Survey Mission for the Chao Phraya Delta, Thailand*, Report to the UNDP, 1968.

[4] "Thailand Not Growing Enough", *Investor*, May 1972.

[5] Sansom, R. L., *op. cit.*

[6] *Fertilizer: a Description of the Industrial Sector in Thailand*, ASRCT, Bangkok, 1968.

[7] *1963 Census of Agriculture*, Bangkok.

[8] Chancellor, W. J., *Survey of Tractor Contractor Operations in Thailand and Malaysia*, Agricultural Engineering Dept., Univ. of California at Davis, 1968.

[9] *Thailand Farm Mechanization and Farm Machinery Market*, a Co-ordinated study project by the Ministry of Industry, Board of Investment, Min. of National Development and Ministry of Agriculture, Royal Thai Government and Kasetsart University, Bangkok, together with the Industrial Finance Corporation of Thailand and USOM, 1969.

[10] *Ibid.*

[11] Developing Countries Trade Promotion Staff, U.S. Dept. of Commerce.

CHAPTER FIVE

CO-OPERATION IN AGRICULTURE

Co-operation in Thai agriculture pre-dates the introduction of the Raiffeisen agricultural co-operative movement. In Thailand, as in other peasant societies throughout the world (including pre-industrial Europe), farmers came together for their mutual assistance and security. Traditional forms of co-operation found in Thailand are the reciprocal labour group *(long khaek)* and the rotating credit institution *(len chae)*. In the *long khaek* system the individual may call upon neighbours and kinsmen to help with agricultural tasks, supplying them with food and refreshment. In turn, the individual must be prepared to reciprocate by assisting on the farms of any of his group members, when necessary. The *long khaek* group does not include the whole village, nor is its membership haphazard or arbitrary. Rather a given group represents an alliance of "friends" and is an expression of the micro-political scene at the village level. The system may also be termed a traditional insurance system, since it ensures that the individual does not unnecessarily suffer when misfortune strikes, such as when a farmer becomes sick or senile or a woman villager becomes widowed.

In the *len chae* system members pay in money to a central fund on a periodic basis; then after a pre-determined period the whole sum is handed over to a single member. The process continues so that each member receives the bonus in turn. The *len chae* similarly does not embrace a whole village, but simply a group of friends. There may also be separate groups for men and women. This system, based upon the economic principle of deferred gratification, cultivates the spirit of thrift and co-operativeness among farmers who may be geographically remote from a hank and, moreover, suspicious of it.

Corvée *(phrai)* was also traditional to Thailand in that the local rulers could summon labour for major works such as road building or canal construction. At present, the village headmen, although salaried government officials and elected by and from the villagers as *primus inter pares*, also have the authority to assemble labour for work activities in the village good, such as repairing tracks, cleaning canals, etc.

It is apposite to mention here that most Thai villages have to support a temple with its population of priests, although nunneries are more rare in rural Thailand. Buddhist priests do not engage in productive work, but instead travel the village each morning, begging for sustenance from the villagers. Each villager gives a portion of his daily food, which gift earns spiritual merit. In addition, the village men form work groups for the construction or repair of temples. In certain rare cases in North East Thailand the villages have been reduced to such poverty that they are unable to support the priests and the temples have had to close down. The Buddhist priests in the Thai village are not outsiders; on the contrary, they are often villagers themselves. The Buddhist tradition enjoins on every man to become ordained, so that farmers have the opportunity at any time in their life to forego for a greater or lesser period their temporal trials and responsibilities and enter the priesthood. In addition, a father may send his small sons to become "temple boys", thus gaining spiritual merit, whilst at the same time reducing the number of mouths to feed in the family. In purely economic terms, therefore, it is difficult to make a case for the temple as an onerous burden upon rural resources. The spiritual benefits are of course outside the scope of this study.

Ideally the *long khaek* and the *len chae* systems could provide a bridge to the modern Raiffeisen movement, but there is little evidence of this in Thailand. Instead, traditional forms of co-operation break down under the process of agricultural modernization and the resultant social stratification. Thus large-scale farmers reportedly find it cheaper to hire labour on the open market than to provide for the refreshment and entertainment of village friends in the *long khaek* system, who, being poorer, may make persistent and embarrassing demands upon the wealthier friend. In the resultant landscape of inequality the agricultural co-operative has a difficult task to survive.

The agricultural co-operative movement was set up in Thailand at the suggestion of European advisers, especially the British, and was based upon the Raiffeisen model. LELART suggests that the birth of the co-operative movement was an instrument of Thai nationalism against the dominance of the ethnic Chinese bourgeoisie [1]. The first co-operative, a credit co-operative to eliminate rural indebtedness, was created in 1916 among a group of farmers in Phitsanulok province. It is, however, curious to note that by 1969 this was the only province outside the areas of insurgent activity which did not have any co-operatives. In 1938 Land Co-operatives were created in Chiang Mai, comprising Land Settlement Societies, Land Hire Purchase Societies, Land Tenant Societies and Land Improvement Societies. The same year saw the birth of Marketing Societies. In 1947 the Co-operative Bank was instituted at a national level, whilst in 1959 Production Credit Co-operatives began to appear. These were a joint experiment by the Royal Thai Government and U.S.O.M. and the first was created in 1960 in Pakchong in Nakhon Ratchasima province. There followed Production Credit Co-operatives in Chachoengsao (1960), Rayong (1962) and Khon Kaen (1963).

Table 6. Growth in numbers of co-operatives in Thailand 1919–1975 [2]

Year	Village Credit Co-operatives	Production Credit	Agricultural Co-operatives
1919	26		
1930	129		
1932	191		
1935	562		
1939	1,793		
1940	2,253		
1942	3,359		
1945	2,555		
1946	4,780		
1947	5,379		
1948	6,196		
1949	7,307		
1950	5,658		
1951	7,276		
1952	8,856		
1953	9,583		
1954	9,819		
1955	9,847		
1956	9,856		
1957	9,890		
1958	9,967		
1959	9,985		
1960	9,962	1	
1961	9,950	1	
1962	9,938	2	
1963	9,876	3	
1964	9,838	5	
1965	9,805	7	
1966	9,784	9	
1967	9,744	10	
1968	9,646	10	
1969	8,078	13	
1970	3,213	15	304
1971	213	29	401
1972	93	33	395
1973	43		460
1974			620
1975			597

Commencing in 1969, a merger was initiated between the Village Credit Co-operatives and the Production Credit Co-operatives to create the Agricultural Co-operatives. *Table 6* shows the growth in the numbers of co-operatives between 1919 and 1975. The increase after 1932 is related to the coup d'état which occurred in Thailand on 24th June of that year, the new government of Pridi Phanomyong being more favourably inclined to the co-operative movement. Village Credit Co-operatives increased more or less steadily in number until 1953, but thereafter increased only slowly until their rapid decline with the initiation of the Agricultural Co-operatives.

The geographical spread of the co-operative movement is somewhat uneven, with a concentration occurring in the more prosperous and favoured Central Plain. In other words, co-operatives are most numerous in the areas of maximum credit-worthiness and maximum profitability rather than in areas of greatest indigence.

Map 8 shows the distribution of Land Improvement Co-operatives in Thailand in 1965. There is a very marked concentration in Chai Nat in the Central Plain which has about 33,000 acres associated with this movement. The province was however, the scene of a major multi-purpose co-operative project assisted by Taiwanese experts and no other province approaches anywhere near this concentration. Land Improvement Co-operatives are generally concentrated in the Central Plain and West with their prosperous commercial agriculture, while the South and much of the North are deficient in such schemes. They are, however, of importance in the North East where they form part of a concerted international effort to improve the efficiency of agriculture of this backward but strategic region. It is noticeable that Phitsanulok (Upper Central Plain), the original home of the Thai agricultural co-operative movement, has less than 1,000 acres under land improvement co-operatives.

It is not appropriate to analyze here the Thai co-operative movement, although such an analysis surely needs to be undertaken, or to join in the argument as to whether the Raiffeisen model of agricultural co-operation has any relevance to the Third World. Suffice it to say there are many anecdotes from Thailand of rural misunderstanding of co-operative aims and of abuse and corruption within the movement. Nevertheless, the writers feel that the co-operative can play two useful major roles in the provision of credit and in the marketing of agricultural produce.

Rural indebtedness is common in Thailand, especially in the agriculturally more disadvantaged areas. The pawn shop is a familiar feature of the Thai market town and sometimes the farmer will borrow cash or food from village friends of roughly similar socio-economic status. Such a relationship in which the provision of a reciprocal loan is implicit precludes exploitation and indigence. Often,

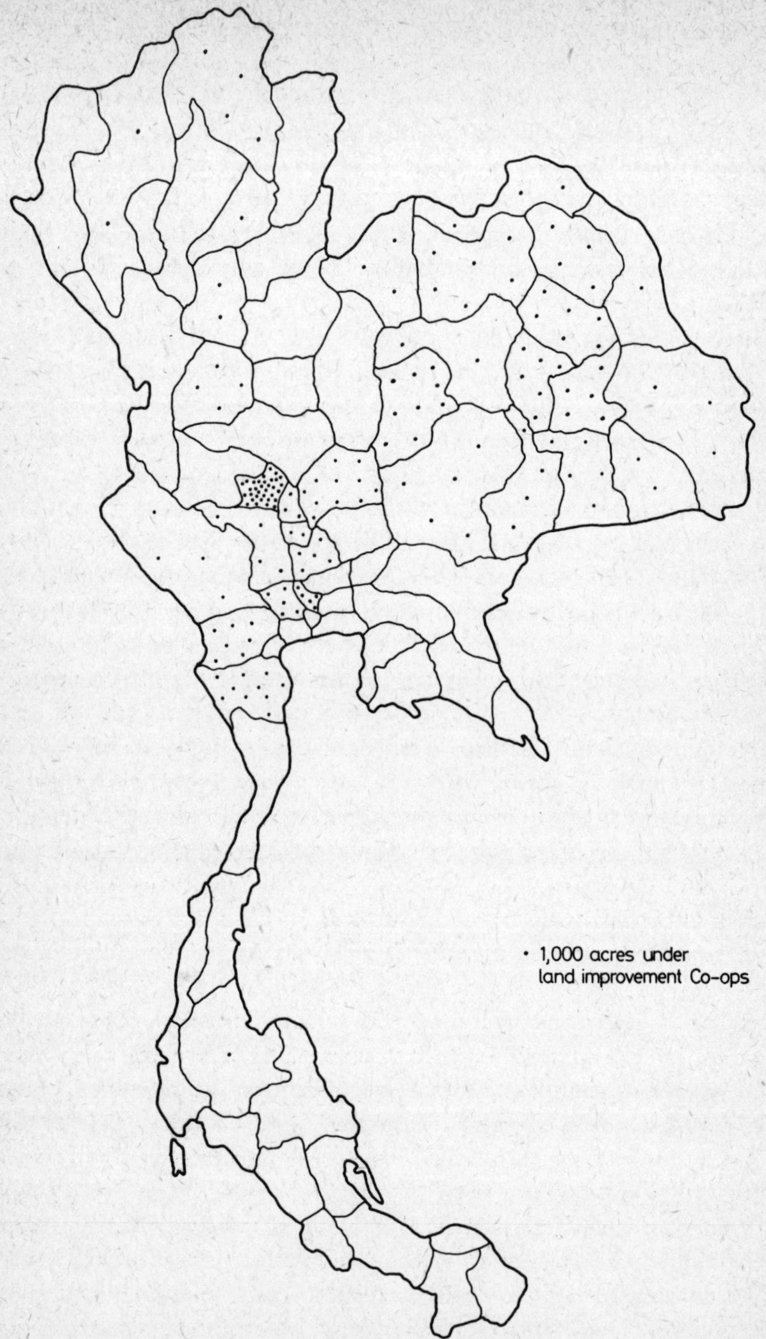

Map 8. Land improvement co-operatives (Source:
Dept. of Land Co-ops., 1965)

however, loans involve the professional moneylender, usually an ethnic Chinese, who may demand high rates of interest. Moreover, loans are often required not for developmental efforts such as the purchase of fertilizer or improved seeds but for basic day-to-day needs—to buy food, medicine, clothes or to educate children. Indeed, O'REILLY observed in North East Thailand that the adoption of agricultural innovations, a goal considered desirable *per se* by both government officials and international experts alike, involved some farmers in economic outlays which left them short of cash for basic necessities or unanticipated misfortunes.

In the marketing of agricultural produce the Thai farmer is often at the mercy of the Chinese middleman. The middleman doubtless aids the evacuation of crops from remote rural areas; indeed, in some cases (if the farmer is desperately short of cash), the middleman will buy the standing crop before harvest and may then employ the farmer as a wage labourer to harvest the crop. But the farmgate price that the farmer receives from the middleman differs markedly (and to the farmer's disadvantage) from the ultimate market price. However, farmers may be unaware of what the price of a given crop should be, since the middleman is their only source of communication. Radio broadcasts of acceptable crop prices have only served to exacerbate tension between the farmer and middleman, since there is no way of compelling the middleman to comply with the recommended price.

In both the provision of credit and the marketing of agricultural produce co-operatives have not yet been able to compete effectively with the Chinese middleman. However, the ubiquitous middleman does play a useful role in the rural economy and is perhaps often unjustly maligned, and it may be neither possible nor desirable to remove him entirely from the rural scene. Instead, the co-operative must become a more efficient and attractive "middleman", whilst legislation can be introduced to prevent the worst excesses of the middleman system.

It should also be remarked that in the Thai village a plethora of developmental agencies may co-exist, including Extension Agents, co-operatives, farmers' groups, Water Users' groups and international experts. Yet there is a lamentable lack of co-ordination and co-operation between these various bodies; there may even be conflicting advice—which may sometimes stem from rivalry and jealousy at the level of the various ministries in Bangkok. At the village level this merely leads to confusion in the minds of farmers and frustrates genuine development.

REFERENCES

[1] Lelart, M., "L'évolution des coopératives agricoles de crédit en Thaïlande", *Archives Internationales de Sociologie de la Coopération*, July–December 1976.

[2] Kochaseni, S., *L'économie thaïlandaise dans l'Asie du Sud-Est*, Thesis of the Faculté de Droit de Toulouse, 1960.

Statistical Yearbook of Thailand

Statistics from Co-operative Promotion Department, Ministry of National Development, collected by Lelart *(op. cit.)*.

CHAPTER SIX

AN EXPORT ECONOMY

Even though Thailand has had an annual growth rate of 7·5% for the last two decades, which is the highest in South East Asia, and has the largest monetary reserves of any non-industrial country [1], nevertheless its economy remains bound to the production and export of a few primary products, as *Table 7* shows.

Table 7. Total value of principal exports, 1970

Commodity	Export value (Baht)	Percentage of total value of exports
Rice	2,526,605,273	17%
Rubber	2,249,678,049	15%
Maize	1,856,912,283	13%
Tin	1,618,522,822	11%
Teak	202,209,731	1%
Total exports	14,772,444,225	

(Source: *Foreign Trade Statistics of Thailand.* December 1970, Department of Customs, Bangkok.)

Of particular importance is the growing percentage of maize in this list. Between 1961 and 1970 the total value of rice as an export decreased by 30%, the total value of rubber and teak as exports increased by 5% and 19%, respectively, whilst tin increased by 162% and maize by 210%.

There are three principal markets for Thai produce. Firstly, rice, poultry, meat and processed foodstuffs find a steady market in Asia, particularly in the increasingly prosperous and industrial economies of Malaysia, Singapore, Hong Kong and Japan, while maize largely goes to a Japanese market. Secondly, the primary products of Thailand including kenaf and cassava find a profitable but uncertain market in the European Economic Community. Thirdly, there is the U.S.A., which although it cannot be categorized with respect to any particular commodity, is a vital importer of Thai produce. Thai exports to Eastern Europe· and the Soviet Union are as yet insignificant, but there is scope for substantial expansion. *Table 8* shows the principal importers of Thai produce in 1970.

Table 8. Principal importers of Thai produce, 1970

Country	f.o.b. value (Baht)	Per cent of total value
Japan	3,770,374,507	26
U.S.A.	1,985,526,305	13
Netherlands	1,276,355,447	9
Hong Kong	1,112,808,480	7
Singapore	1,018,078,553	7
Malaysia	768,173,348	5
Taiwan	719,539,123	5
Federal Republic of Germany	533,289,805	4
Indonesia	341,647,482	2
Saudi Arabia	322,453,062	2

(Source: *Foreign Trade Statistics of Thailand*, December 1970, Department of Customs, Bangkok.)

(Note: 50 Baht is approximately equal to £1 sterling.)

As far as new field crops are concerned, maize is the most significant with a total export of 1,371,474 metric tons in 1970, 47% of which went to Japan and 32% to Taiwan. Also important is kenaf with a total export of 253,905,693 kg and a total value of 708,856,641 Baht, the principal importers being Japan (27%), Belgium (8%), France (6%), West Germany (2%), Italy (4%), and Spain (5%). Cassava totalled 1,176,074,128 kg with a value of 1,009,762,434 Baht. The principal component of this is cassava flour, 52% going to Canada, 36% to Laos and 3% to Hong Kong, although cassava meal and cassava pellets are also exported. Groundnuts and groundnut cake exports totalled 11,292,066 kg with a value of 10,716,109 Baht, Hong Kong taking 39%, Malaysia 15% and Singapore 15% of the total. Soya beans and soya bean cakes totalled 6,762,624 kg with a value of 12,279,821 Baht, the principal importers being Malaysia (60%) and Singapore (26%). Kapok is dwindling in importance as an export, but totalled 11,286,612 kg with a value of 17,677,250 Baht, almost all of which went to Japan [2].

These then are the principal non-rice crops which are helping diversify the traditional rice-export economy of Thailand. With the exception of kenaf and cassava they are largely dependent upon a fairly restricted Far Eastern market, in particular upon demand in Taiwan, Japan, Hong Kong, Singapore and Malaysia. However, it is unlikely that these countries will be able to achieve agricultural self-sufficiency or in fact will attempt to achieve it, and the future for Thai products in these markets is therefore quite promising.

The export of livestock and livestock products is one which is at present relatively small, but could increase in importance. The total export of water buffaloes was 35,299 head in 1970 which earned 73,425,287 Baht. The major importers were Hong Kong, Malaysia and Singapore. Pigs totalled 15,533 head and earned 10,774,304 Baht, Hong Kong and Laos being the chief importers, while live poultry, exported to Hong Kong, Malaysia, Singapore, Indonesia and Japan, totalled 4,535,301 head and earned 838,388 Baht. Fresh, chilled or frozen poultry went mainly to Japan, totalling 1,256 kg and earning 70,536 Baht [3]. The prospects for expansion in the export of live animals seem to be fairly limited, but there could well be room for expansion for beef, poultry and poultry products.

The basis of the Thai economy has remained remarkably uniform for over a century, although at present it is more diverse than at any previous time. *Table 9* shows the importance of exports from Siam in 1850.

Table 9.
Siam's exports around 1850

Commodity	Value (Baht)
Raw cotton	450,000
Dried meat	120,000
Rice	150,000
Pepper	99,000
Tobacco	100,000
Sugar	708,000
Total	4,331,000

(Source: Malloch, D. E., *Siam, Some General Remarks on its Productions*, Calcutta, 1852.)

Of particular interest is the relatively great importance of sugar and raw cotton at that time. In fact, contemporary observers foresaw the Siamese economy becoming increasingly dependent upon sugar cane. However, both sugar cane and cotton have since waned in importance, although in recent years there has been some new expansion in both.

Although at an early date rice was being exported from Siam as a royal monopoly, it was the Bowring Treaty of 1855 which first opened up the country to international trade and provided a stimulus to rice production. From 1857–1859 to 1900–1904 the annual export of rice rose from 990,000 piculs to 11,130,000 piculs and thereafter grew steadily to 26,290,000 piculs in 1951 (1 picul equals 60 kg) [4].

The significance of this brief historical survey is that it shows that at some time in the past a major transformation took place in Thai agriculture, whereby the area under and the production of rice increased rapidly to meet the export demands of a monetary economy. Although the government of the day gave encouragement and assistance to this activity, most of the initiative came from the farming class and the enterprise was wholly Thai rather than Chinese or European. History, therefore, lends confidence to the hopes that the agriculture of Thailand may in the future be equally flexible and malleable in the face of major economic and political changes in the world at large.

At present Thailand, like all developing countries, is seeking to industrialize. Thus during the period from 1960 to 1965 the annual growth rate of primary commodity exports was 8·1%, whilst the rate for manufactured and semi-manufactured exports was 11%. However, if these manufactures are analyzed, it is apparent that many of them consist of the processing of local raw materials for a large number of export industries, especially tin, tapioca, flour and timber. Secondly, there is the export of products developed in connection with import substitutes, particularly Portland cement and petroleum products [5]. It seems, nevertheless, that the industrial base of Thailand will continue to be bound to the agricultural economy and will be unable to expand without the prior development of the agricultural sector. Moreover, in view of Thailand's relative paucity of mineral and energy resources, it seems certain that the mainstay of the economy will continue to be agriculture. Therefore, Thailand cannot afford to neglect its agriculture either in matters of investment or research.

REFERENCES

[1] Press Release No. 27, Permanent Mission of Thailand to U.N., New York, January 23rd, 1968.
[2] *Foreign Trade Statistics of Thailand*, Dept. of Customs, Bangkok, December, 1970.
[3] *Ibid*.
[4] Ingram, J. C., *Economic Change in Thailand, 1850–1970*, Stanford University Press, 1971.
[5] *Foreign Trade Statistics of Thailand, op. cit.*

CHAPTER SEVEN

A RICE ECONOMY

Rice is the staple of the people throughout the Kingdom and there is no province in which rice is not grown. The bulk of this rice is paddy rice, cultivated in bunded fields, inundated by rainfall, by flooding of streams or lakes or by irrigation. Upland rice or dry rice, on the other hand, is cultivated on land that is either too high or too steep for inundation to take place (even though the degree of slope or elevation may be only slight). Naturally, it depends on rainfall for its moisture. Upland rice may be grown as a sideline by paddy cultivators on their upland fields, while in hill areas, where it is the only rice cultivable, it provides a staple crop for the Hill Tribes.

Map 9 shows the distribution of upland rice cultivation in Thailand. Paradoxically, this map does not include the Hill Tribe cultivators, since they are unenumerated for census purposes and the difficult terrain they occupy and their hostility to central government has so far effectively barred any thorough survey. Upland rice is predominantly concentrated in Loei in the North East, where there are about 36,000 acres, while Nan in the North also has a large amount. Concentrations are observable throughout the North, North East and peninsula, but upland rice is noticeable for its absence in the level Lower Central Plain.

Paddy rice in Thailand may be further subdivided into glutinous and non-glutinous varieties. Glutinous rice differs from the non-glutinous variety in that the endosperm and pollen grain of the former have glutinous starch, whilst in the case of the latter they contain common starch. The two types, glutinous and non-glutinous, are distinguished in the Thai nomenclature as "sticky rice" and "lordly rice", respectively. The non-glutinous variety is produced in most of the rice-growing areas of the world, while glutinous rice is a more localized phenomenon within South Eastern Asia, providing the staple of North and North East Thailand, Laos and parts of Burma.

Map 10 shows the distribution of glutinous rice cultivation in Thailand. The highest absolute amounts are found in Chiang Rai, Udon Thani, Sakon Nakhon, Kalasin, Roi Et and Ubon Ratchathani, each of which has over 375,000 acres. The second category (250,000 to 375,000 acres) comprises the provinces of Nong Khai, Nakhon Phanom and Khon Kaen, all in the North East. The third category (125,000 to 250,000 acres) comprises Chaiyaphum, Maha Sarakham and Si Sa Ket

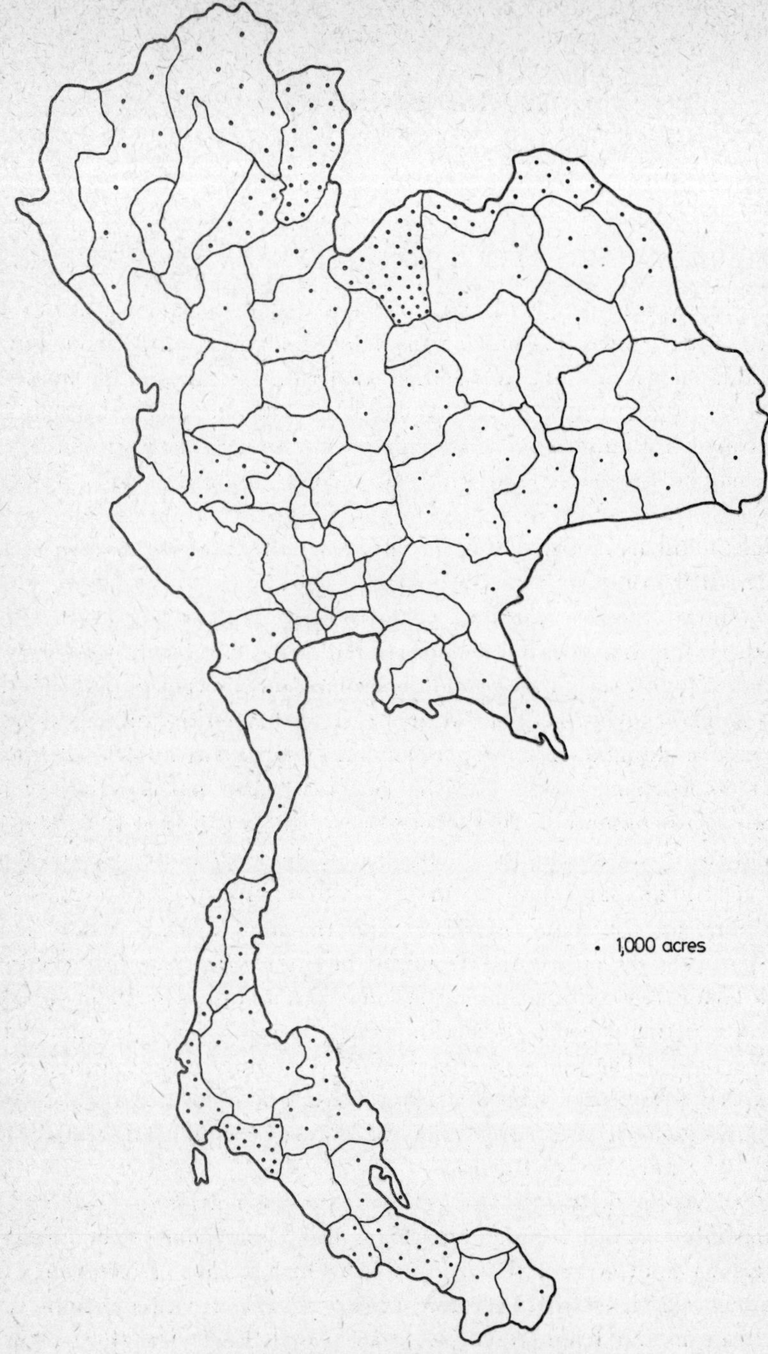

Map 9. Upland rice cultivation (Source: Statistics of Upland Crops and Vegetables, 1970)

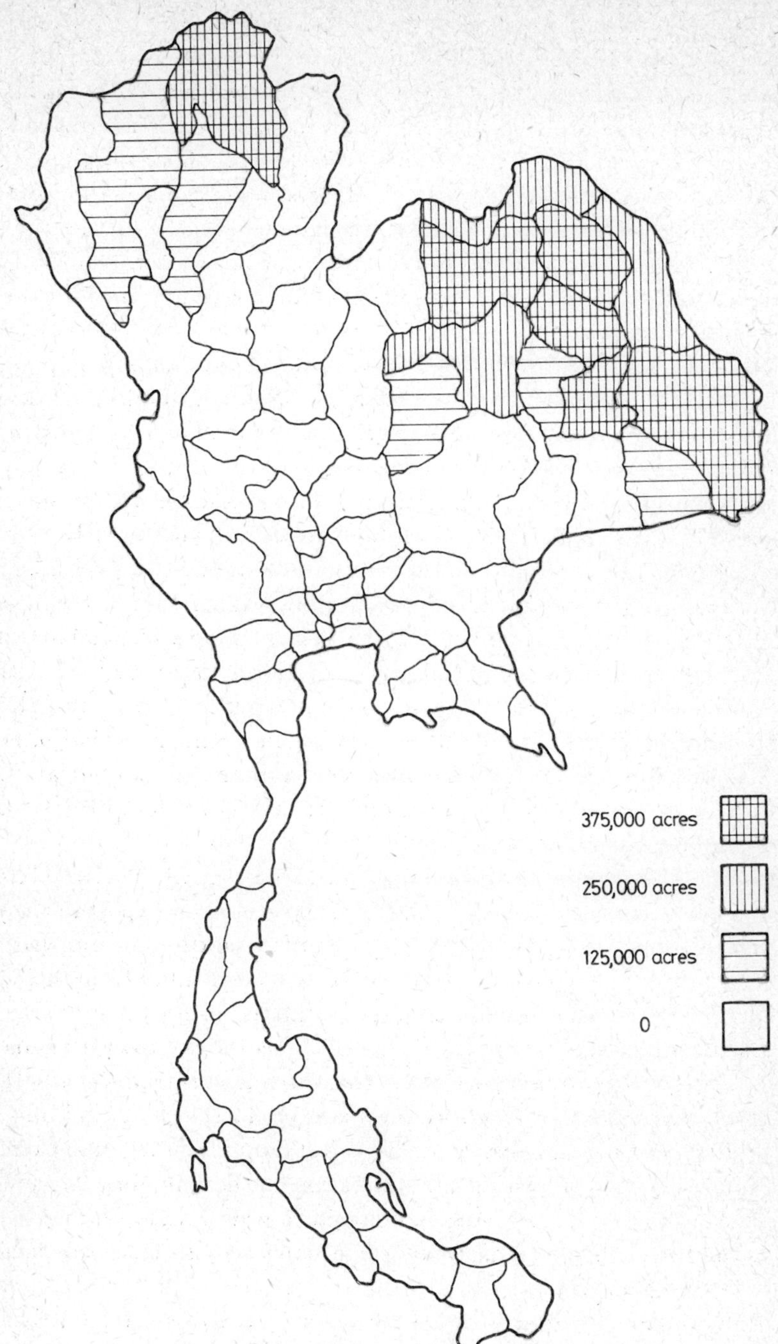

Map 10. Glutinous rice cultivation (Source: Rice Report, 1975)

in the North East and Chiang Mai and Lampang in the North. Glutinous rice is scarcely cultivated at all in the Central Plain and South — where it is found it is for confectionery and alcoholic beverages rather than for staple consumption. However, glutinous rice is also unimportant in the North Eastern provinces of Surin, Buri Ram, Nakhon Ratchasima and Loei. In Loei the importance of upland rice and a traditionally mixed farming economy is the reason. In the case of Surin and Buri Ram it is explained by the Khmer influence on these provinces, while in the case of Nakhon Ratchasima the answer lies in the growing commercialization of this the gateway to the North East and the most advanced North Eastern province.

Map 11, showing the distribution of non-glutinous rice cultivation, presents the reverse picture. Totals are low in the North and North East, although some non-glutinous rice is grown in these areas traditionally as a luxury foodstuff or for use in confectionery or alcohol, but increasingly as a commercial crop. This is especially marked in the North Eastern province of Nakhon Ratchasima, which has over 375,000 acres under non-glutinous rice. Surin and Si Sa Ket provinces with their populations of mixed Thai-Khmer descent have between 250,000 and 375,000 acres. High totals are also found in Phra Nakhon Si Ayutthaya, Suphan Buri, Nakhon Sawan and Phichit in the Central Plain, in Prachin Buri and Chachoengsao in the East and in Nakhon Si Thamarrat in peninsular Thailand. All these provinces have over 375,000 acres under non-glutinous rice cultivation. The second category (250,000 to 375,000 acres) includes Sara Buri, Pathum Thani, Chai Nat, Kamphaeng Phet, Phitsanulok and Sukhothai in the Central Plain and Nakhon Nayok in the East. The third group (125,000 to 250,000 acres) is scattered through all regions except the North.

Rice is both a commercial and a staple crop. Traditionally, Thailand has been a major rice producer and exporter and rice remains a major export to the present day, which places the Thai farmer in a favourable position in comparison with farmers in other developing countries whose cash crops differ from their staples. Fluctuations in the world price of rice cannot ruin the farmer, since the rice can be consumed locally. The Central Plain is *par excellence* the rice bowl of Thailand, the major producer of rice for export, while the North East is traditionally the most subsistent rice region, although its total rice production is second only to the Central Plain and in some years exceeds it. But despite this regional pattern, the subsistence–commercial ratio in rice cultivation for the individual farmer is likely to vary from season to season depending upon a wide variety of socio-economic factors, and it is therefore misleading and dangerous to label rice farmers as subsistent or commercial.

Although some of the countries of Monsoon Asia that constituted Thailand's principal market for rice are now seeking self-sufficiency in their own production,

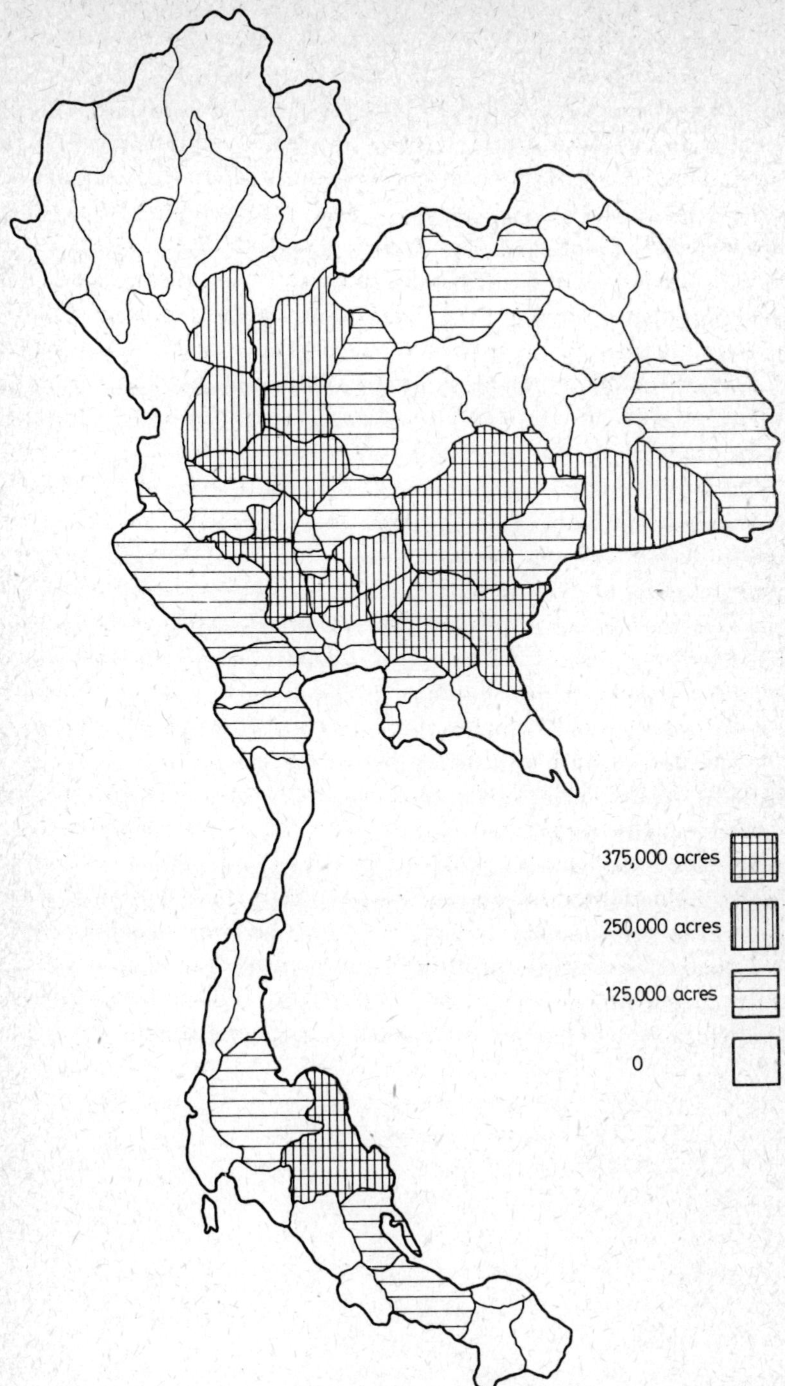

Map 11. Non-glutinous rice cultivation (Source:
Rice Report, 1975)

it is unlikely that rice exports will collapse, since world demand for this almost universally acceptable food item is growing. Obviously, the future for non-glutinous rice is the more favourable and the commercial potential of glutinous rice will remain strictly limited. It sustains the rural populations of the North East, and provides a surplus for the non-agricultural and urban populations of these regions (many of whom are increasing their consumption of non-glutinous rice at the expense of glutinous rice). It also provides a surplus for a relatively small export market, which consisted mainly of Laos (before the political upheavals), but also includes Japan, Malaysia, Hong Kong and Singapore. It is probable, then, that the future will see an increasing expansion in non-glutinous rice cultivation throughout Thailand at the expense of the glutinous variety.

Improved strains of rice are an important innovation in Thailand as throughout the rest of the rice-cultivating areas of Monsoon Asia, and at present improved verieties account for about 30% to 40% of the total area. A great deal of attention has been given to the so-called "miracle rices", which are able to treble existing yields. These varieties, IR5, IR8, IR20 and IR22, were produced at the International Rice Research Institute at Los Banos, Philippines. At present, however such miracle rices are not of great importance in Thailand, partly because of the lack of a domestic market for the somewhat acquired taste of miracle rice, but just as significantly because these strains require good water control, which is not yet available throughout Thailand.

The future of rice must also be viewed in the perspective of the growing importance of upland crops in the economy of the country. These dry-season crops do not normally compete with paddy rice for land, since their requirements are different; they do, however, compete for labour and investment. Certainly, upland crop production has been given a major fillip by the extension of irrigation facilities making dry-season production of upland crops possible. Whether Thailand should concentrate on its wet-season paddy production or its dry-season upland crop production is a controversial issue. It seems to the writers that it is a prudent policy to have as diversified an agricultural base as possible.

CHAPTER EIGHT

UPLAND CROPS

Upland crops, sometimes called field crops, are crops which are grown on so-called upland, or land that because of its relief (slight though it may be) is unsuited for wet-rice cultivation. They are usually cash crops. Strictly speaking, therefore, upland rice is by definition an upland crop, but because of its economic role it seemed more logical to treat it in Chapter 7. A threefold land-use classification is recognized in Thai agriculture by both scholar and peasant alike. The three categories are a) paddy *(na)*, b) upland *(rai)*, and c) orchard or garden *(suan)*. *Suan* crops will be dealt with in Chapter 9. This traditional distinction is sometimes less clear-cut now, since commercial motives may induce farmers to plant upland cash crops even in their paddy fields, whilst similar motives may lead to the expansion of the fruit and vegetables of the *suan*, traditionally for domestic consumption, onto the *rai* or *na*, where they are cultivated on a larger scale as cash crops.

Essentially for cash, upland crops may be relatively new in the country or province and may have required the clearing of new land from the forest. There is a lack of stability about upland crop cultivation which is not characteristic of rice cultivation, and innovations may be adopted quite rapidly in an area through favourable prices and/or extension efforts, but may be rejected just as rapidly when prices drop. It is also in the study of upland crops that one sees the greatest regional diversity and even the very *raison d'être* of an attempt to delimit agricultural regions within Thailand. Moreover, the picture is far from static. The introduction of new crops, their geographical diffusion, and the simultaneous decline and contraction of other crop areas lend a dynamism to the regionalization of Thai agriculture which it is impossible to capture by the delimitation of agricultural regions.

The government published Statistics of Upland Crops and Vegetables list annually a wide variety of crops, as shown in *Figure 1*. Although all are of interest and all are considered in the determination of agricultural regions, some are too insignificant for mapping. Accordingly, only the most important ones have been selected, for discussion here.

Maize (*Zea mays* Linn.) is by far the most important upland crop in Thailand, whose production, according to BEHRMAN followed the opening up of areas free

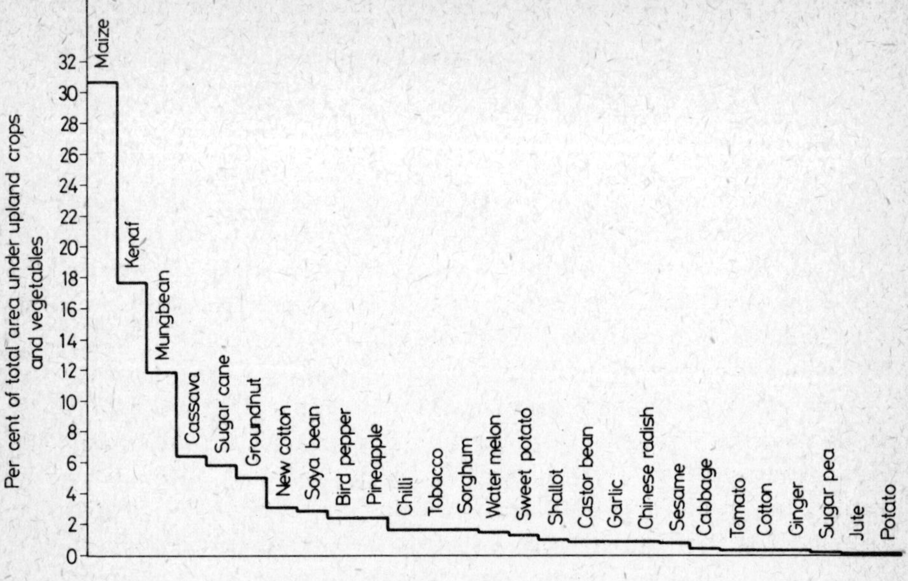

Fig. 1. Cultivation of Upland Crops and Vegetables in Thailand, 1970 (Source: Statistics of Upland Crops and Vegetables)

from malaria [1]. It is chiefly concentrated in a bloc of provinces in the Upper Central Plain with Nakhon Sawan as the core where the limestone content of the soil is considered to be favourably high. Maize is unimportant in the Metropolitan area and the contiguous southern Central Plain and in the peninsula as *Map 12* shows, and is relatively unimportant in the North and in the interior of the North East. However, Nakhon Ratchasima, the gateway to the North East, has about 207,000 acres under the crop and it is likely that it will spread into the North East.

Kenaf is a crop which has assumed considerable importance in the economy and is the single most important innovation to have gained a footing in the agricultural economy of North East Thailand. ZIMMERMAN in his survey in 1932 made no mention of its presence or potential [2]. Kenaf is the name used in the trade for the fibre obtained from two closely related species of the family Malvaceae, *Hibiscus cannabinus* L. and *Hibiscus sabdariffa* L. var. *altissima*. Thai and Chinese varieties have grown wild for centuries, whilst the Cuban variety was introduced by U.S.O.M. in 1951. Whilst SATO states that the Cuban type has almost entirely replaced the native and Chinese varieties [3], BEHRMAN claims that the native type still predominates [4]. Kenaf as a cash crop spread rapidly in

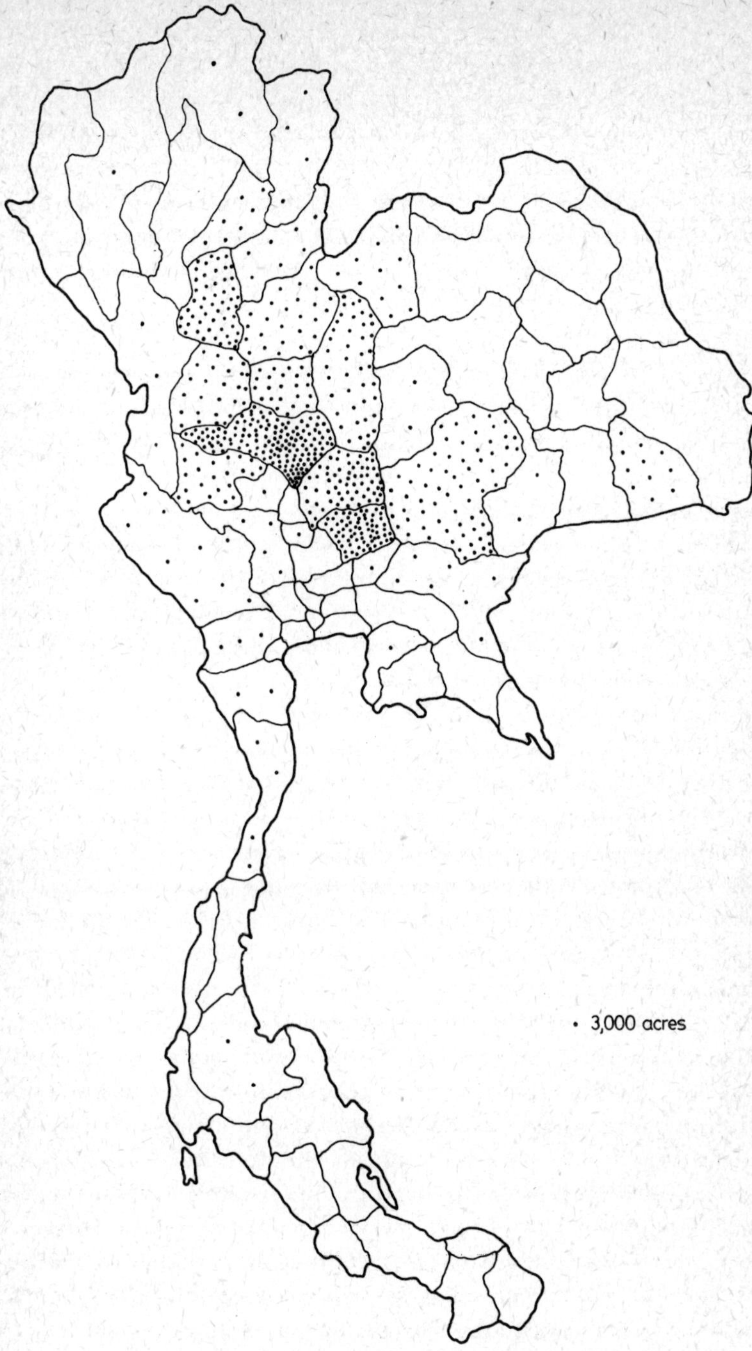

Map 12. Maize cultivation (Source: Statistics of
Upland Crops and Vegetables, 1970)

the North East, being suited to that region's peculiar edaphic and climatic characteristics and in 1961 many farmers actually shifted from maize to kenaf cultivation. Kenaf is primarily an export crop, the fibre being used in the manufacture of sacks, but as it is inherently inferior to jute its production is very sensitive to economic conditions in Bangladesh and North East India. Bad years in Bangladesh, through natural disasters or political upheavals, have been years of high farmgate prices for Thai kenaf. The farmgate price has, however, fluctuated alarmingly from year to year, producing considerable economic distress as well as some prosperity among growers.

Map 13 shows kenaf to be a North Eastern crop *par excellence*. It is cultivated on a significant scale in all North Eastern provinces with the exception of Nong Khai, but the core area is clearly Maha Sarakham. In only one non-North Eastern province is kenaf cultivation of a significant scale, viz. Nakhon Sawan in the Upper Central Plain.

The core of mung-bean (*Phaseolus aureus* Roxh.) cultivation, is the Upper Central Plain with particular concentrations in Nakhon Sawan and Sukhothai *Map 14*. Outliers exist in Phetchaburi and Ratchaburi in the West, in Chon Buri in the East and in Nakhon Si Thammarat and Phatthalung in the peninsula. Nakhon Ratchasima has about 6,000 acres under the crop, and it is likely that it will spread throughout the North East.

Cassava (*Manihot utilissima* or *Manihot esculenta*) was first introduced into Thailand from Malaya as an intercrop between young rubber trees in the South. An improved strain was introduced into the Kingdom by the Japanese sometime between 1914 and 1945, which was processed by small plants in the Chon Buri area until the American company, Thai Tapioca Ltd., set up an establishment in Rayong and planted 2,000 acres to the crop. The innovation was soon adopted by local farmers and demand for cattle feed in Europe led to the production of dried chips [5]. Although the Eastern provinces of Chon Buri, Rayong and Chachoengsao remain the core area of cultivation, the distribution of the crop is nevertheless diffuse *(Map 15)*. It is scattered throught the Central Plain and West, but is largely absent from the North. Nakhon Ratchasima has a high concentration of cultivation, but it is also significant in Chaiyaphum, Khon Kaen, Udon Thani, Kalasin and Nakhon Phanom in the North East. Recent observation suggest a further spread of this crop into North East Thailand [6]. The crop is primarily grown as an export crop for cattle feed for the European Economic Community but it also forms a component of the rural diet, although the writers are unaware of cassava being consumed as a staple, as it is in parts of Africa, South America and in Indonesia, where among indigent groups there has been a shift away from rice to the cheaper and more filling cassava. It should also

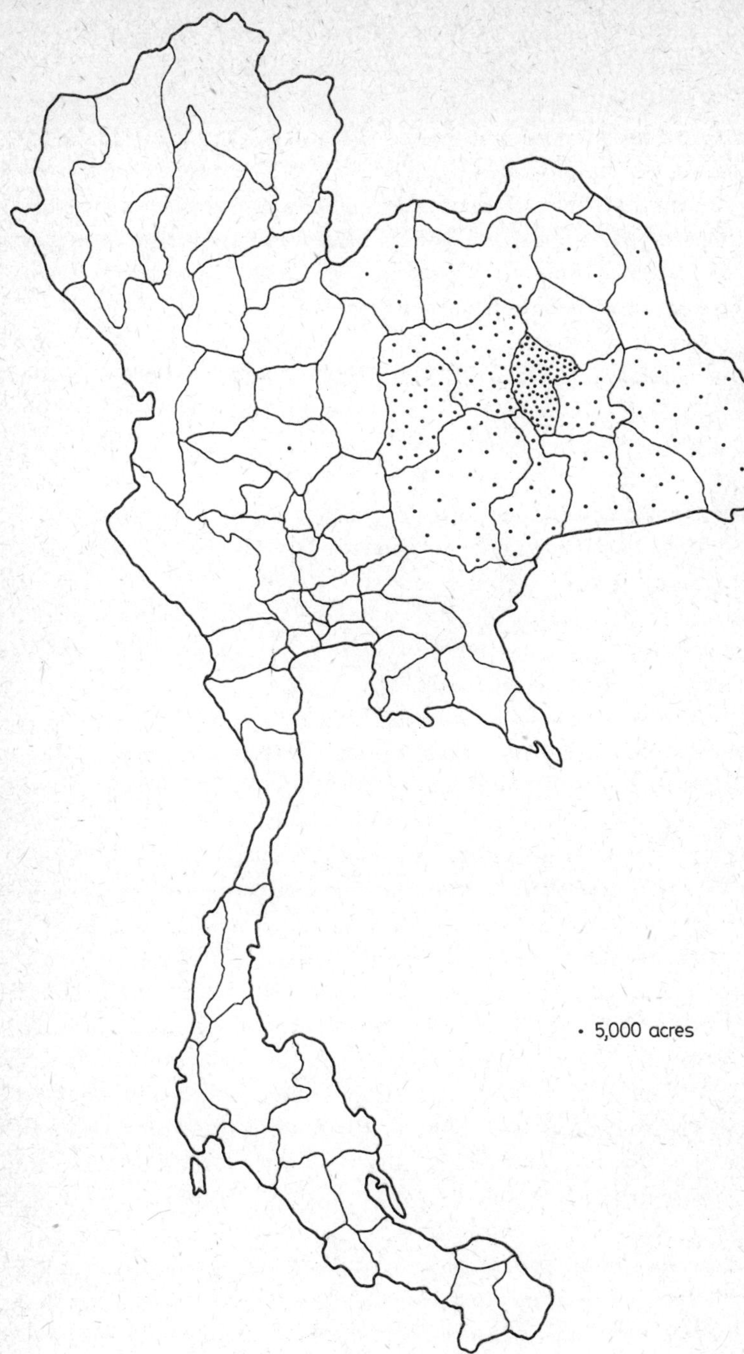

Map 13. Kenaf cultivation (Source: Statistics of
Upland Crops and Vegetables, 1970)

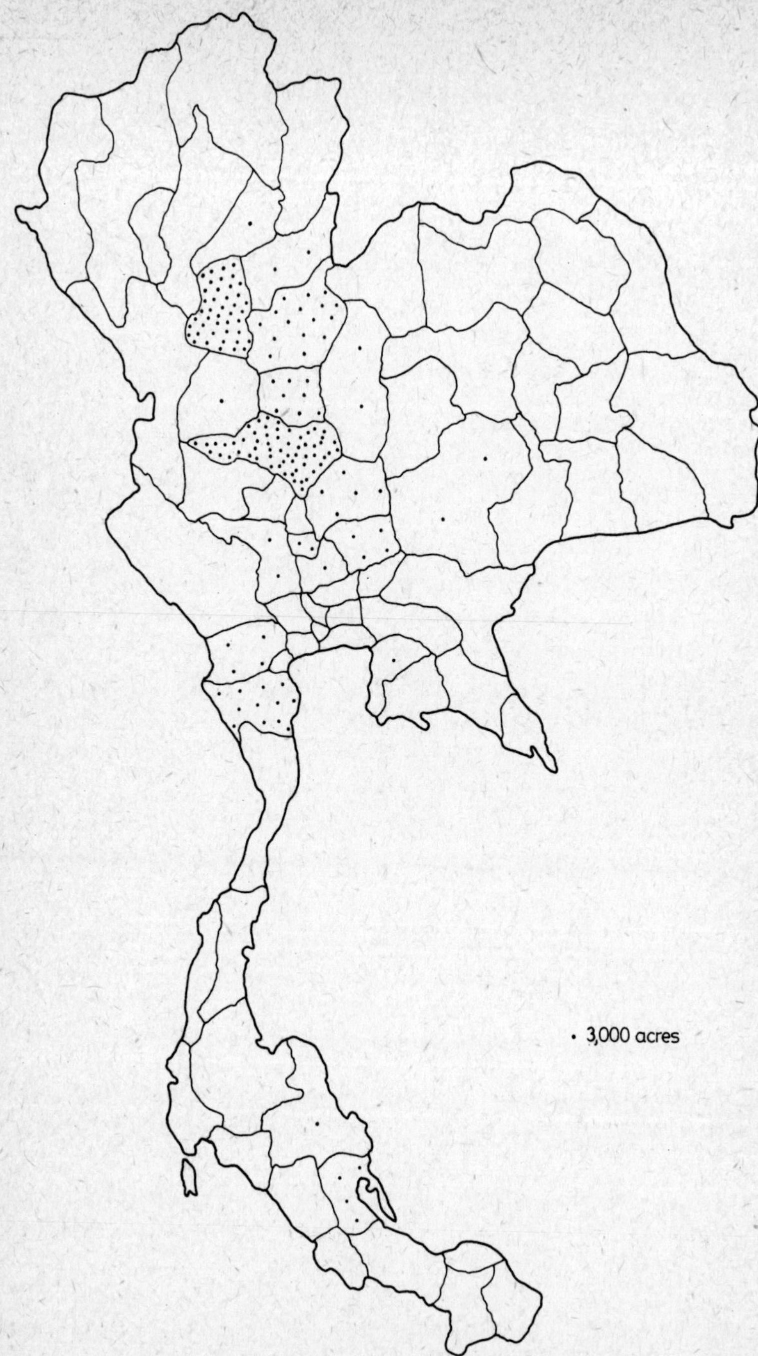

Map 14. Mung bean cultivation (Source : Statistics of Upland crops and Vegetables, 1970)

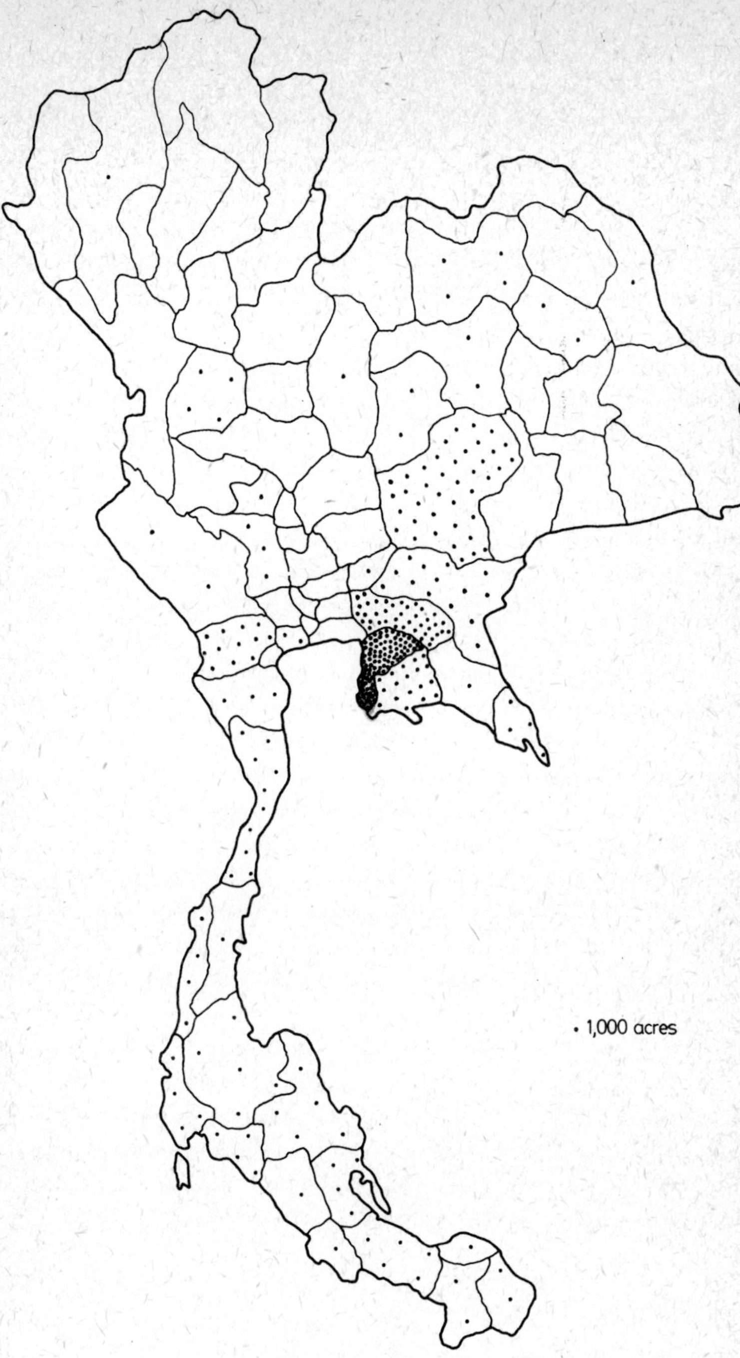

Map 15. Cassava cultivation (Source: Statistics of
Upland Crops and Vegetables, 1970)

be added that monoculture of the crop has a deleterious effect, particularly in areas of poor soil structure.

Map 16 shows the distribution of sugar-cane (*Saccharum officinarum* Linn.) cultivation, which is concentrated in the Lower Central Plain and West with Suphan Buri having a particularly high acreage. Although not significant in the South, the crop is cultivated fairly widely throughout the Upper Central Plain, North and in parts of the North East. Udon Thani in the North-North East has a particularly high acreage, presumably related to recent developments in large-scale irrigation in that province.

The distribution of groundnuts (*Arachis hypogaea* Linn.) is very diffuse *(Map 17)*. They are cultivated throughout the Kingdom and most provinces have at least 1,000 acres under the crop. The only areas where it is insignificant are the Metropolitan and coastal provinces of the Central Plain, extending as far inland as Sing Buri. Notable concentration of cultivation are found in Sukhothai (Upper Central Plain), Nakhon Sawan (Upper Central Plain) and Nakhon Ratchasima (Near North East). Groundnuts, in addition to their use as human food (they are particularly popular in the Far East), can be used as livestock fodder. They are exported throughout the Far East.

New cotton (*Gossypium* Spp.) is treated separately from local cotton, since it warrants a separate heading in the Annual Statistics of Upland Crops and Vegetables. It is found in all regions except the South but is especially concentrated in Sukhothai in the Upper Central Plain *(Map 18)*. There is heavy concentration also in Nan in the North and in Loei in the North—North East. British technical aid was furnished to help the development of cotton production and an experimental cotton farm was set up in Tak Far in the Upper Central Plain. Local weaving of clothing, a traditional skill throughout rural Thailand, is gradually giving way to the mass-produced factory product and it is hoped that the increased cotton acreage together with a skilled and artistic workforce may be able to produce clothing not only for the home, but also for the overseas market.

Map 19 shows the cultivation of soya beans (*Glycine max* L. Merrill) (Leguminosae) or *Glycine soja* Sieb. & Zucc. (Leguminosae). Absent from the South and from the North East except for Nakhon Ratchasima, it is scattered throughout the North and Central Plain but with heavier concentrations in the North. Sukhothai in the Upper Central Plain emerges as the chief area of concentration with about 118,000 acres under the crop.

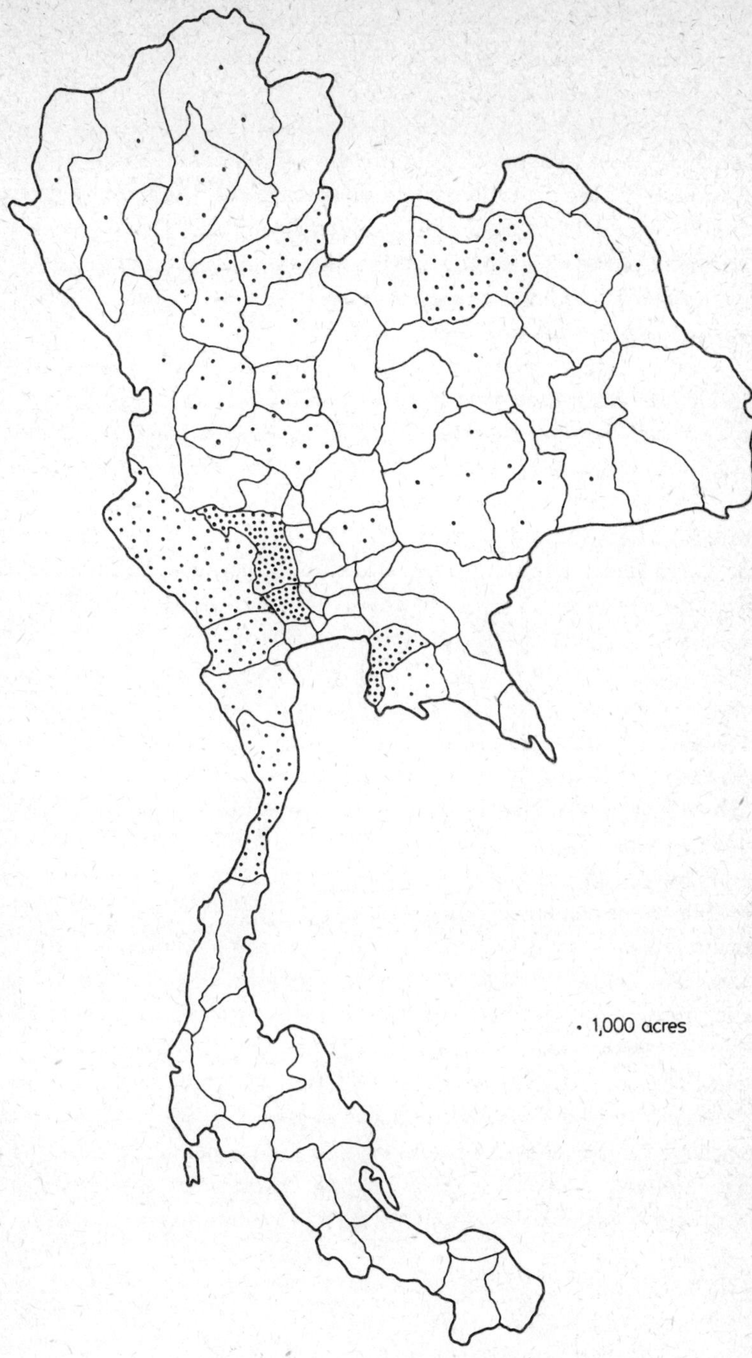

Map 16. Sugar cane cultivation (Source: Statistics of
Upland Crops and Vegetables, 1970)

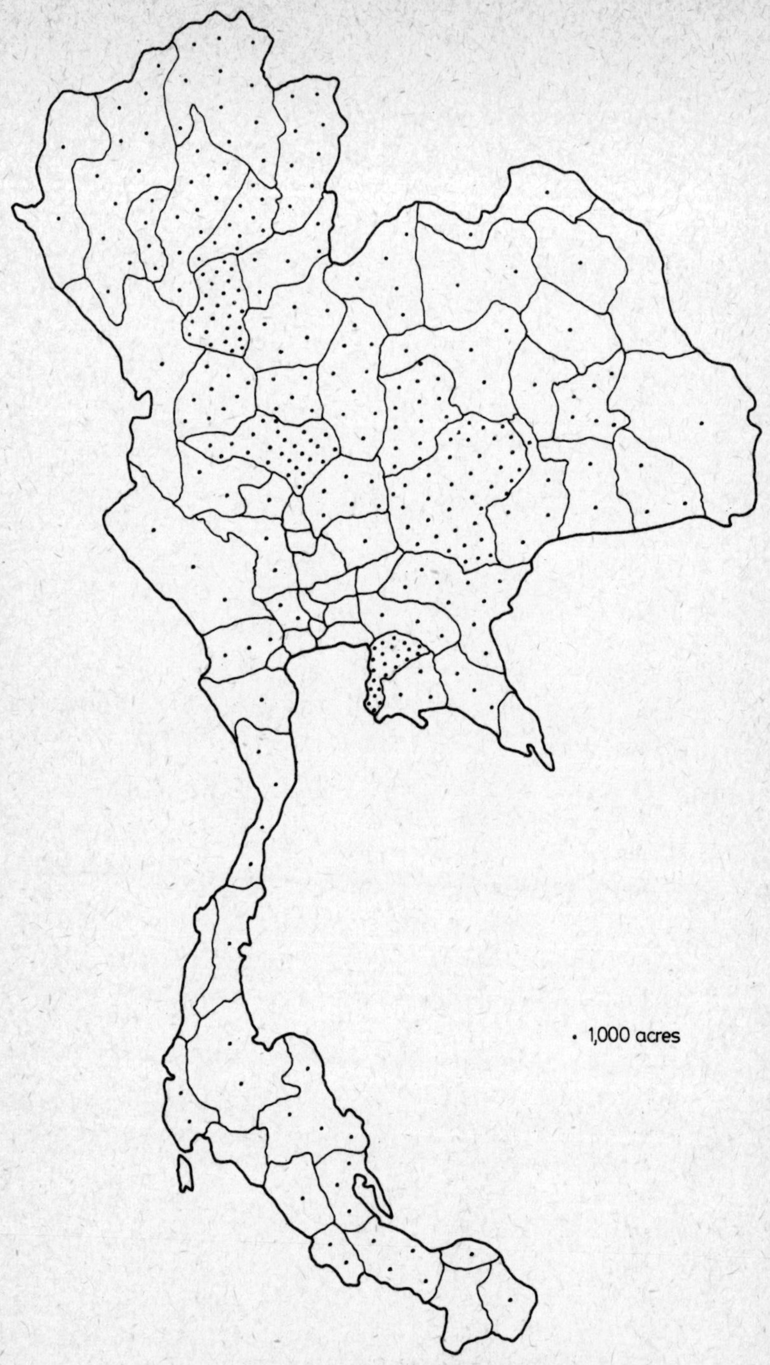

Map 17. Groundnut cultivation (Source: Statistics
of Upland Crops and Vegetables, 1970)

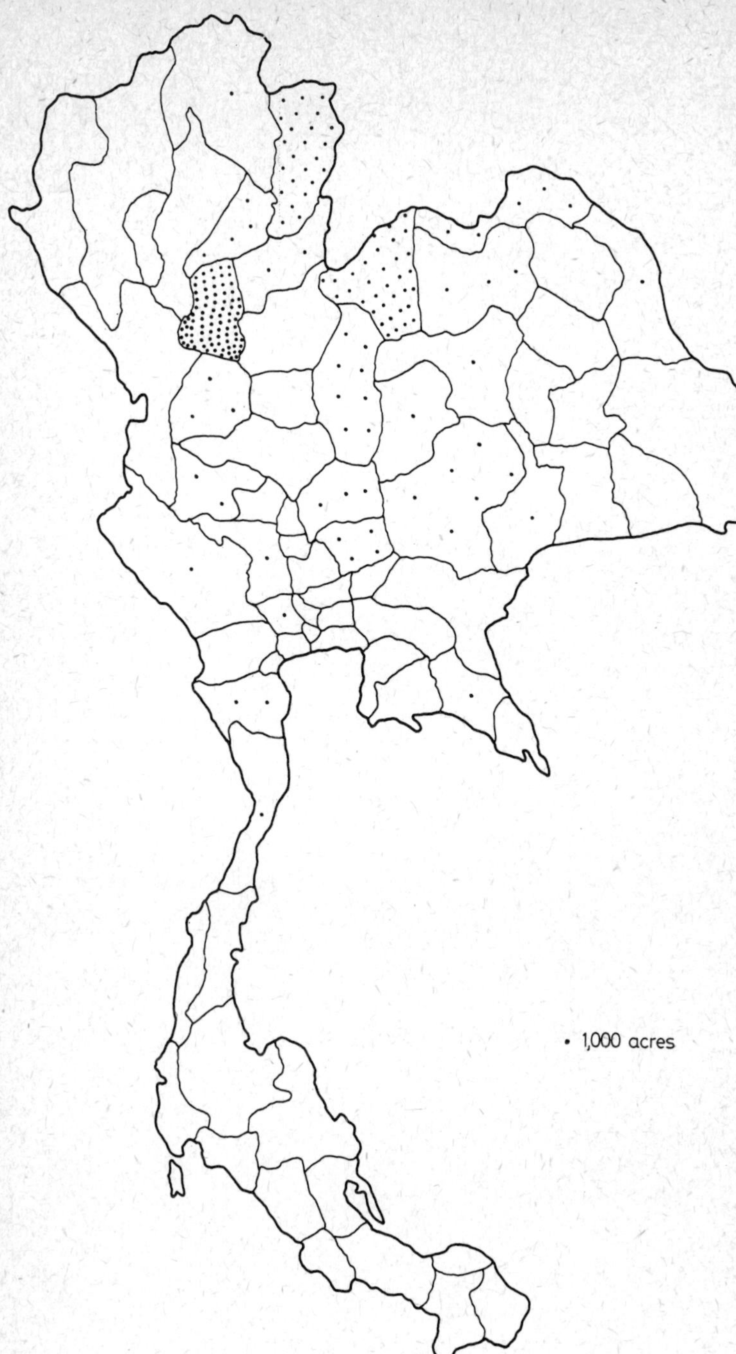

Map 18. New cotton cultivation (Source: Statistics
of Upland Crops and Vegetables, 1970)

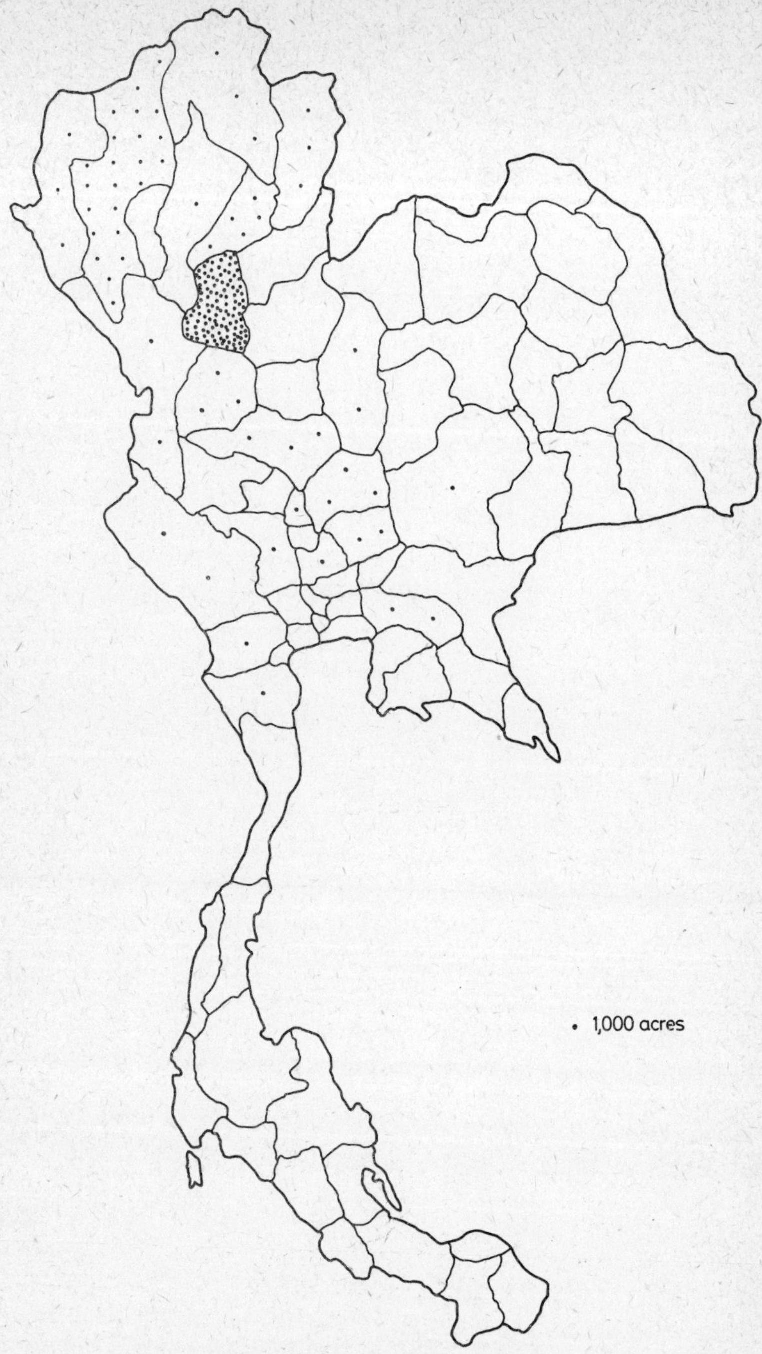

Map 19. Soya bean cultivation (Source: Statistics of
Upland Crops and Vegetables, 1970)

REFERENCES

[1] Behrman, J. R., "The Significance of Intracountry Variations for Asian Agricultural Prospects", *Asian Survey*, Vol. 8, 1968.

[2] Zimmerman, C. C., *Siam: Rural Economic Survey 1930–'31*, Bangkok Times Press 1931.

[3] Sato, T., *Field Crops in Thailand*, Centre for South East Asian Studies, Kyoto University, Japan, Series N–1, 1966.

[4] Behrman, J. R., *op. cit.*

[5] Mr. Peter Kirrage of Thai Tapioca Ltd., Bangkok (personal communication).

[6] Khun Pittaya Hiranburana, Enumerator of Royal Irrigation Department, Bangkok (personal communication).

CHAPTER NINE

RUBBER

Rubber was first planted in Thailand during the first decade of the 20th century. It was introduced from Malaya as a substitute crop for pepper, because the price for pepper was declining. Moreover its introduction was entirely due to Thai initiative; the seeds were imported from Malaya in 1901 by one Phya Radasadnupradit. This contrasts with the rubber industries of other important producers such as Malaysia, Indonesia, Sri Lanka, Nigeria and Liberia, which owe their industries to Western initiative.

In southernmost Thailand rubber is the principal crop, contributing an average of 11·6% of the South's Gross Domestic Product between 1960 and 1969 [1]. Since the development of foreign plantations was not encouraged by the Thai government it is chiefly a smallholder crop. Moreover, in Narathiwat, Yala and Pattani many of the cultivators are Moslem Malays, whilst Chinese settlers established holdings in Songkhla and Yala.

Rubber development in Thailand proceeded unguided by the government until the 1930s when a passive interest developed with the introduction and execution of the Rubber Control and Restrictions Act (1934). In 1959 the government, realizing the importance of the crop, approved the Rubber Replanting Aid Fund Act, whilst the Rubber Research Centre was established at Hat Yai in 1965 [2]. Int the same year Thailand also started to receive international assistance through the UNDP/FAO Rubber Development Project. The distribution of rubber in Thailand is a particularly regionally specific crop *(Map 20)*. There is a small bloc of cultivation in the Eastern provinces of Trat, Chanthaburi and Rayong, but the main concentration is in the peninsula, commencing in Chumphon and reaching its highest concentration in the extreme South in Songkhla and in Yala and Pattani on the Malaysian border. Spatially speaking, this area is merely a northward extension of the much larger and more productive Malaysian rubber area. Rubber is not cultivated at all in the other provinces of the Kingdom, although some processing factories are actually anomalously located in Nakhon Phanom, 700 km northeast of Bangkok.

Rubber cultivation represents a long-term investment to both the peasant and the nation. It is not a seasonal crop like maize which may be planted one year and rejected the next, but a tree crop which takes many years to mature and reach

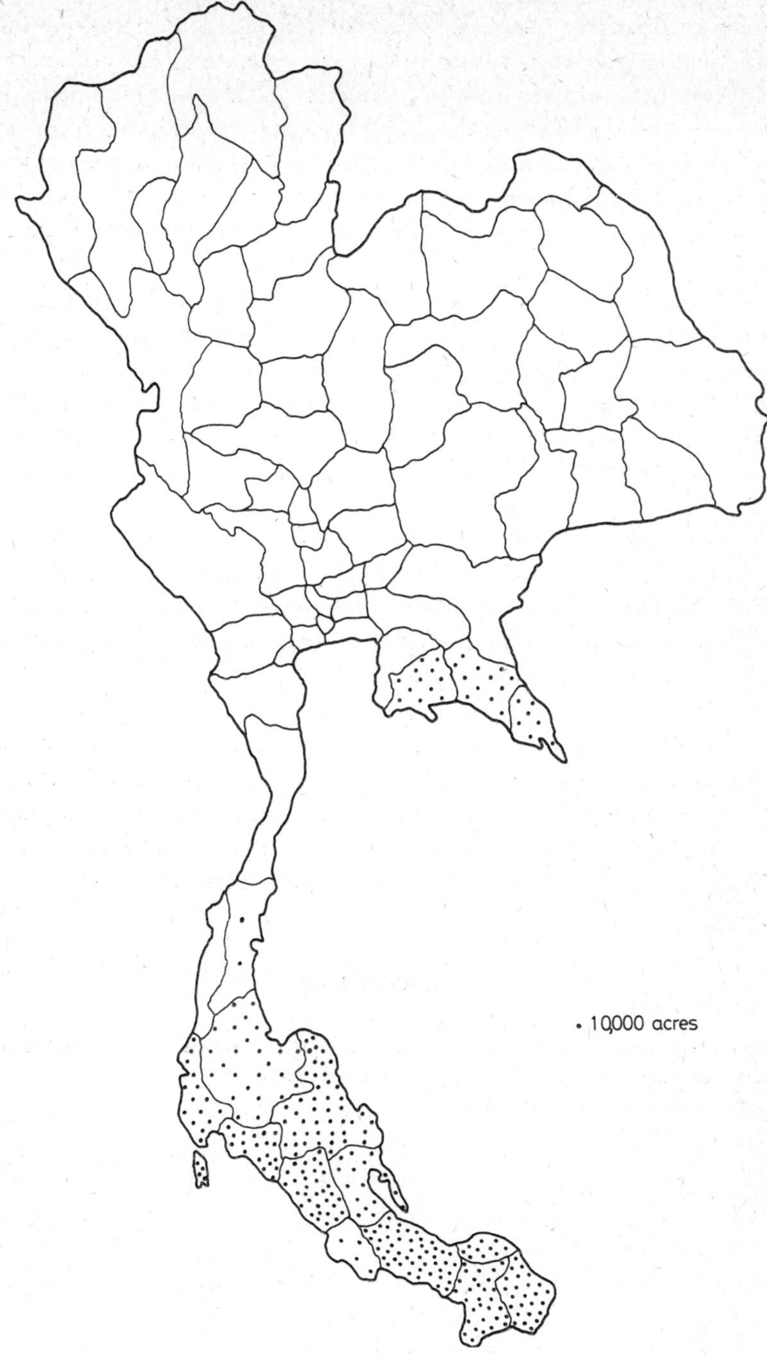

Map 20. Rubber cultivation (Source: Rubber
Research Centre, 1971)

maximum productivity. Thus it is estimated that Thai rubber trees produce their highest yield in the 16th to 18th year after planting, but although yields thereafter decline, the trees are still productive in their 54th year after planting [3]. For this reason and also because of rubber's importance as a foreign exchange earner it is likely that the rubber-growing area will remain a permanent feature of Thailand's geography. Expansion into new provinces is unlikely: rather there may well be some contraction and consolidation in view of the problems that rubber now faces.

The chief problem is competition from synthetic rubber, which hits the Malaysian industry hard, but the Thai industry even harder, because Thai rubber is inferior and lower-yielding than the Malaysian product. Then there is the ageing of the trees and consequent decline in yields. Finally there is the politically sensitive nature of the core of the rubber producing provinces. Not only are these provinces the home of Malay Moslems, who did not wish to join Thailand and have a history of unfortunate and strained relations with the Thai authorities, but these provinces also harbour the remnants of the Communist forces of the Malayan Emergency, who fled to the Thai border after their defeat by British forces. Recently they have reportedly been joined by disaffected students and intellectuals following the ending of Parliamentary democracy and the imposition of military rule in Thailand in 1976.

Just as Malaysia is diversifying its base away from rubber, notably to oil palm, so also it is in Thailand's interest to diversify, for although the country as a whole is not dangerously dependent upon rubber, certain provinces certainly are. Moreover, they are provinces distinct in religion and ethnicity and plagued by disaffection and insurgency. Hence any slump in the world market for rubber could lead to greater dissatisfaction and upheaval in these sensitive border provinces.

REFERENCES

[1] *A Survey of Rubber Growing Areas of Thailand*, Rubber Research Centre, Hat Yai, 1971.
[2] *Rubber Development Project Phase II*, FAO, Rome, 1974.
[3] Rubber Research Centre, Hat Yai.

CHAPTER TEN

FRUITS AND VEGETABLES

A wide variety of fruit and vegetables are grown by farmers all over the Kingdom in their *suan* or gardens. Primarily, this produce supplements the diet but any excess is sold directly, probably by the farmer's wife, in the market or by the roadside. Both rural and urban dwellers are especially partial to fruit and even in poor rural areas consumption seems relatively high. Although most of this small-scale production of fruit would escape the census, some farmers do experiment with fruit and vegetables on a larger scale. For example, when irrigation water became available in Kalasin province some farmers began to cultivate water melons as a dry season crop on their paddies after the rice harvest.

A more recent development is Thailand's entry into the canning of tropical fruits and vegetables and not only has this development stimulated peasant production in favourable areas but in some cases the canning companies have established their own large plantations. For example, the Universal Food Company Ltd. has two plantations, each of over 790 acres, one in Lampang and the other in Chiang Mai province. Their canned produce includes pineapples, mangoes (for juice), rambutan, cucumbers and papaya. Fully 60% of the produce is supplied by local farmers. Another example is the Siam Food Products Co. Ltd. of Bangkok which has a 2,372 acre plantation in Chon Buri province, producing pineapples, rambutan and baby corn. Local farmers supplement the production of this plantation. Again the Siam Food Supply Co. Ltd. of Samut Prakan grows fruit and vegetables on its vast plantations in Chon Buri and Prachuap Khiri Khan provinces, while the Thai Pineapple Industry Corporation Ltd. is also based in Prachuap Khiri-Khan.

The Tropical Fruits and Vegetables Co. Ltd. has its factory in Chon Buri and owns extensive plantations. Three plantations in Chon Buri, comprising 1,545 acres under cultivation, produce pineapples, asparagus, and tomatoes. A fourth plantation in Prachin Buri is devoted to growing bamboo shoots, an Oriental delicacy, while canned rambutan, a speciality, is purchased exclusively from farmers in Chanthaburi province.

Bamboo occurs freely in Thailand but the best quality is found in Kanchanaburi province. The Siam Pan Trading Co. Ltd. has a production plant there which produces first quality bamboo cane for export [1].

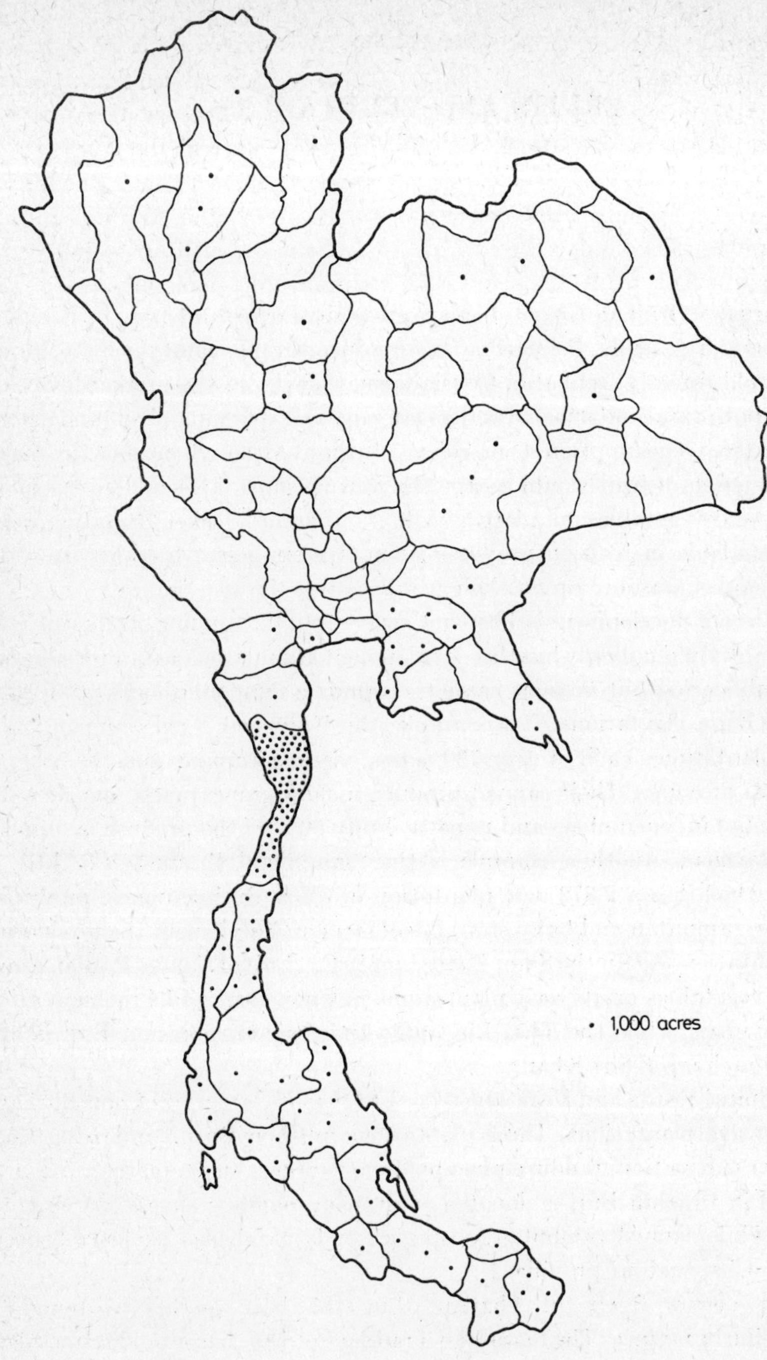

Map 21. Pineapple cultivation (Source: Statistics of
Upland Crops and Vegetables, 1970)

Without doubt pineapple is the most successful fruit export, 14,519 metric tons of tinned pineapples being exported in 1973 [2], but other canned fruits such as longan and rambutan are less successful. Possibly this is because pineapple has great appeal in the industrialized Western countries, whereas the more exotic-seeming fruit have an exclusively Oriental market. However, Thai canned fruits face severe competition in the Western countries from other tropical or sub-tropical producers, including the U.S.A., Australia and South Africa with their highly efficient fruit plantations and their stringently hygienic processing. The future prospects of Thai canned fruits are a matter of speculation.

Map 21 shows the distribution of pineapple cultivation in Thailand. Production in Prachuap Khiri Khan far exceeds that of any other province. There is also a definite concentration of cultivation in the peninsular area, but elsewhere cultivation is scattered throughout the various provinces of the Central Plain, North and North East.

A vide variety of vegetables and condiments are grown as *suan* crops, including cucumbers, long cucumbers, garlic, shalotts, onions, water melons, parsley, tomatoes, eggplants, sugar peas, potatoes, sweet potatoes, ginger, cabbage and Chinese cabbage, Chinese radishes, chilli and bird peppers. Not all these crops, however, are of sufficient importance to be included in the census. The table below shows the acreages under the most important of these *suan* crops in 1970 *(Table 10)*.

Table 10. Acreages under important suan crops in
1970

Name	Acreage
Bird peppers (*Capsicum frutescens*)	156,830
Chilli (*Capsicum annuum*)	101,960
Water melons (*Citrullus lanatus*)	90,781
Sweet potatoes (*Ipomoea batatas*)	77,567
Shallots (*Allium ascalonicum*)	60,359
Garlic (*Allium sativum*)	57,188
Chinese radishes (*Raphanus sativus*)	55,992
Cabbage (*Brassica oleracea*)	27,416

Table 10. (cont.)

Name	Acreage
Tomatoes	18,186
(*Lycopersicon lycopersicum*)	
Ginger	12,609
(*Zingiber officinale*)	
Sugar peas	8,558
(*Pisum sativum*)	
Potatoes	2,084
(*Solanum tuberosum*)	

(Source: Statistics of Upland Crops and Vegetables 1970.)

—Bird peppers are widely grown throughout the Kingdom, but there is a marked concentration in Prachuap Khiri Khan *(Map 22)*. Cultivation is also heavy in Ratchaburi and, surprisingly, Loei in the North East. Although most garden vegetables find a purely local market, bird peppers and chillies assume some importance as export crops, and in 1971 Thailand accounted for 10% of the world export of both these commodities. The principal world markets consist of Sri Lanka (28%), West Malaysia (14%), the U.S.A. (23%) and West Germany (16%) [3]. There is scope for further expansion in production and for the further development of industries processing peppers and chillies to produce the sauces and pastes beloved of the Oriental palate, which are also becoming more popular in Western countries with their increasingly cosmopolitan culinary tastes.

REFERENCES

[1] *Thai Exports*, Thai Export Directory Ltd., Bangkok, 1973.
[2] *Ibid.*
[3] *Plantation Crops: A Review*, Commonwealth Secretariat, London, 1973.

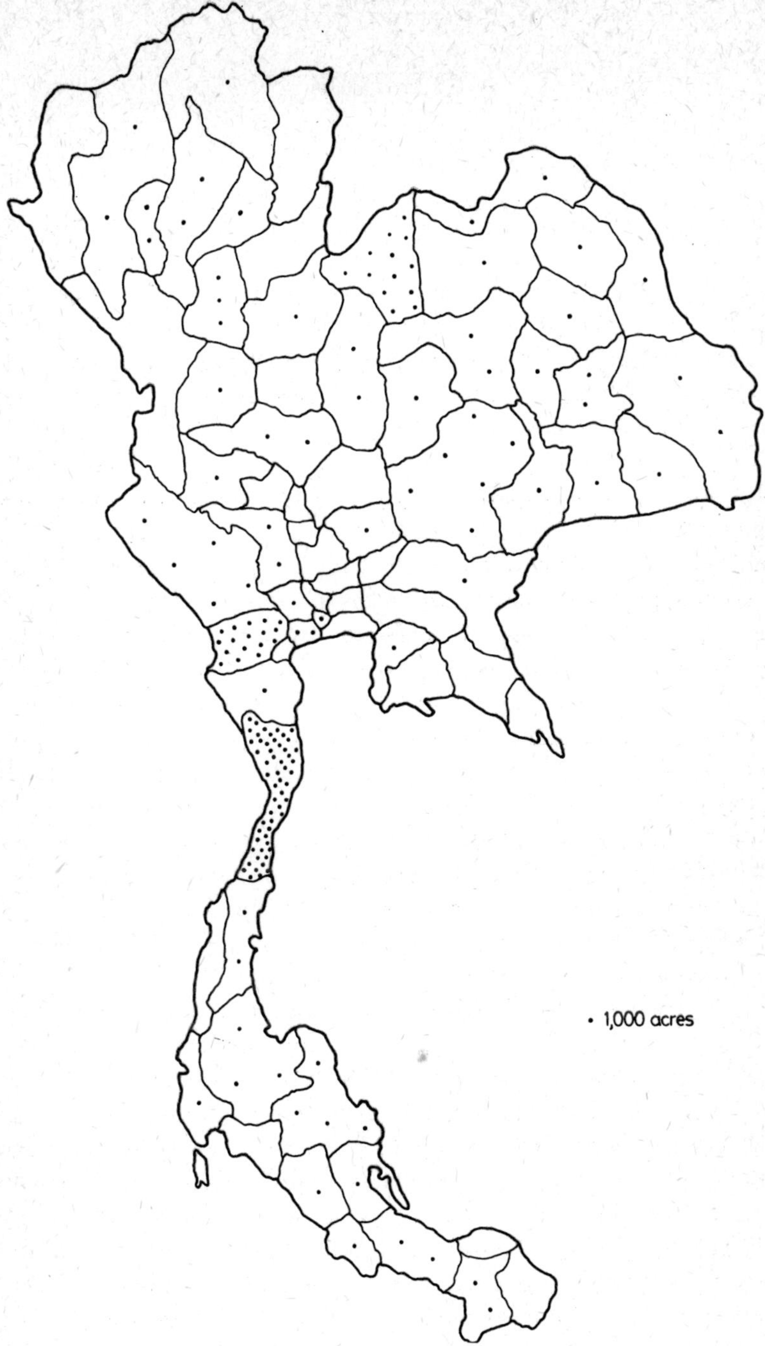

Map 22. Bird pepper cultivation (Source: Statistics
of Upland Crops and Vegetables, 1970)

CHAPTER ELEVEN

LIVESTOCK

Livestock occupy an important role in the economy of the Thai village. Every farm has poultry and chickens and eggs sold in the market provide useful supplements to the farmer's income. Where there are appropriate bodies of water, ducks are kept and, in fact, some farmers have been misusing the state irrigation canals for this purpose. Such farmers interviewed by the writers considered duck raising a more profitable use of the water than the irrigation of crops. Duck's feet are a Thai delicacy and hearsay has it that the demand has even led to imports from English farms.

Water buffaloes are ubiquitous in Thailand. They are essentially beasts of burden, used for ploughing the wet paddy lands. Cattle are rather less common as work beasts, their role being the ploughing of dry upland fields where water buffaloes would be unsuitable. Both types of animal are liable to be slaughtered for meat when they outlive their working life. Some farmers breed and keep more cattle and/or water buffaloes than they use or need, and act as suppliers to their neighbours. In addition, the North East region as a whole is an area of cattle and water buffalo surplus and provides work beasts for the farmers of the Central Plain. North Eastern beasts are also transported by rail to the Bangkok abattoirs where they supply much of the urban market for meat and the growing demand for beef in the many restaurants and hotels in Bangkok and other cities. The growing influx of Western tourists inflates this demand, as some Thai have a prejudice against the over-consumption of beef, since they believe it leads to serious illness in old age. The meat from both water buffalo and cattle appear to be equally acceptable in Bangkok's expensive restaurants.

Thailand also exports cattle products and in 1972, for example, sinews, tendons, parings of raw and waste hides and skins were sent abroad to the value of 8,415,182 Baht, while in 1973 cattle bones and horns to the value of over 50 million Baht were exported. There is certainly scope for expansion in cattle and to a lesser extent water buffalo production. Yet, at present, conditions on the farms are far from ideal. There is almost no improved pasture and instead animals browse the stubble of the rice fields, road verges and voluntary vegetation or forage in the woodland. The individual farmer cannot keep more than a relatively small number of animals which are normally tethered underneath his house because numbers are insufficient to justify the construction of pens. North Eastern farmers expressed to the writers their feeling that lack of security prevented them from investing too heavily in cattle and water buffaloes. Thefts of beasts are relatively common and large-scale cattle rustling is a problem in the region, the rustlers apparently being politically disaffected insurgents.

Other problems facing the cattle industry are poor breeding practices, lack of quality control and the fact that there is no price variation for different cuts and quality. Moreover, the law prohibiting the moving of slaughtered beef (and pork) across provincial lines results in long cattle drives, causing weight loss and the risk of death and disease, as well as a reduction in the price to the producer.

Nevertheless, there are examples of large-scale capital-intensive ranches such as the Chokchai Ranch in Nakhon Ratchasima. The herd was built up around a prize Santa Gertrudis bull and 372 Santa Gertrudis calves imported from the U.S.A. Pastures were improved by sowing with Mauritius grass and Guinea grass, whilst Napier grass was sowed for silage [2]. The Chokchai Ranch is outstandingly successful. Hopefully such ranches will by instruction and animal sale help raise the level of livestock enterprise among local peasant farmers.

Dairy produce is not very important in Thailand, because the Thai like the Chinese did not traditionally consume milk and dairy products, although soya bean milk is widely consumed. However, with the growing Western presence in the form of tourists and businessmen and with the increasing Westernization of the Thai elite a demand for dairy products has been established. This is partly met by the Thai-Danish dairy farm in Saraburi province and the Thai-German dairy farm in Chiang Mai province, as well as by small-scale Indian urban dairy farmers. However, at present the cost of processing local milk is higher than imported powdered milk [3].

Pigs are also a common feature of the Thai village except in the Moslem South and pork is an important component of Thai cuisine. Although every farmer has cattle and/or water buffaloes, pigs are not so numerous, since they are not an essential part of the productive activity. Normally farmers keep a few pigs beneath the house, where they consume household refuse. In addition, they are allowed to forage in the forest. The three established breeds of pig, the Red from the North East, the Kwai from the North and the Hainan from the Central Plain are being supplemented by imported Hampshires, Berkshires, Landrace, Duroc-Jersey and Large White, although it appears that indiscriminate cross-breeding is diluting the characteristics of the indigenous breeds.

The mapping of livestock distributions is no easy matter despite excellent government annual livestock reports. In the first place there is absolutely no information on the poultry population and it would be unreasonable and perhaps unrewarding to attempt to enumerate them. Moreover, most poultry, unlike large beasts, do not enter the larger commercial milieu — they are simply consumed on the farm or sold in the local market. Suffice it to say that chickens are very numerous and ubiquitous in rural Thailand, and regional differences in their distribution on traditional farms are unlikely. Concentrations of ducks, on the other hand would be expected near water bodies, both natural and man-made.

There is evidence, however, that large-scale commercial enterprises, mostly in the Central Plain, may effectively price the smaller producer out of the poultry business. These may have flocks of several thousand chickens, one enterprise near Bangkok having over one million birds, often bred under "battery" conditions [4].

Private American capital has also been invested in poultry raising and processing in Thailand. For example, Arbor Acres (Thailand) Ltd., a branch of the worldwide American company, supply chickens to poultry farmers (as well as to farmers in other countries in South East Asia) and then repurchase the eggs and broiler chickens at a guaranteed price. The broilers are sold in pre-packed form on the Thai and Far Eastern market, while the eggs are exported, especially to Hong Kong [5].

It is most disappointing that no statistics are available for the number of cattle, water buffaloes and pigs per province. Although annual figures are published for the numbers of deceased, diseased and vaccinated animals, one can hardly use these to postulate animal populations and the only reasonable figure to use are the number of animals slaughtered per province. This is not completely satisfactory, primarily because many beasts are sent to the abattoirs of Bangkok for slaughtering. Phra Nakhon province thus emerges with an abnormally high number of slaughtered beasts even though its actual population of beasts is very low. Excluding Phra Nakhon from the distribution therefore provides a more meaningful regional distribution of animals slaughtered. In rural areas animals may be slaughtered in a number of ways. They may be slaughtered on the farm, but this is either uncommon or clandestine, since Buddhism forbids the slaughtering of animals. Traditionally, therefore, the Buddhist farmer takes his beasts to the town, where the cattle are slaughtered by Moslem Indians and the pigs by Catholic Vietnamese, although increasingly provincial capitals are building modern abattoirs. On the journey from the farm to the local place of slaughter it is of course possible for beasts to cross procincial boundaries. For this reason also then the available data are not entirely satisfactory but must suffice.

Map 23 shows the distribution of slaughtered water buffalo in the Kingdom. Although the North Eastern provinces, the main areas of water buffalo are relatively important, the lower Central Plain provinces, the recipient areas for North Eastern water buffaloes, are equally important. Ratchaburi province on the Burmese border is the high spot.

78

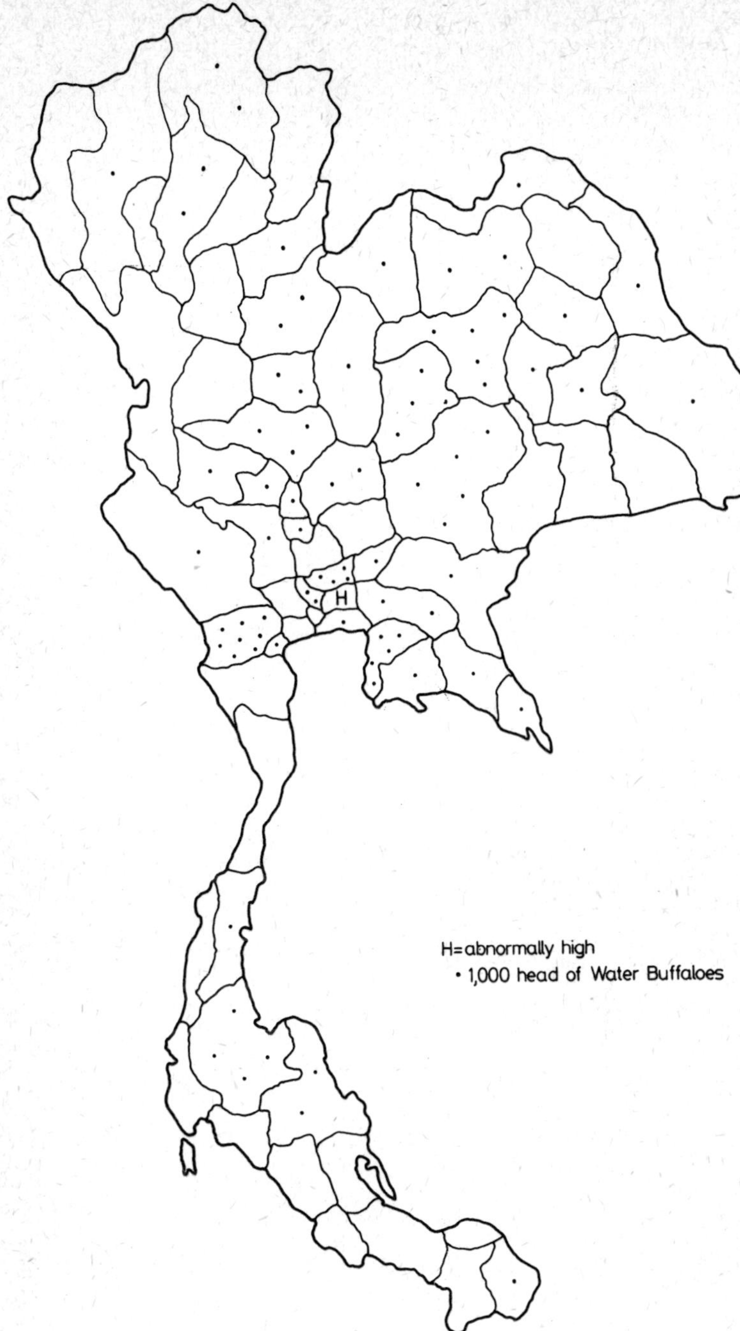

Map 23. **Water buffaloes slaughtered (Source: Livestock Report, 1970)**

The distribution of slaughtered cattle is fairly even, but there are very low scores in Chumphon, Phangnga, Krabi, Phuket and Surat Thani *(Map 24)*. There are heavy concentrations in the North East, Lower Central Plain and North with Ratchaburi again scoring highest. The high concentration in the North is partly a reflection of the large demand in the city of Chiang Mai, an important centre of commerce and tourism. The surprisingly high concentration in the South may also be due to the presence of the urban areas of Hat Yai and Songhkla, but may also be a response to demand from across the border in Malaysia.

Although the number of pigs slaughtered is again high for Ratchaburi, it is Nakhon Pathom which scores highest *(Map 25)*. There is, however, a fairly even distribution throughout the Kingdom, even in the Moslem South, although in this area pigs are raised and consumed by Thai Buddhists and members of the Chinese minority. The large urban markets of Malaysia with their substantial Chinese populations may also stimulate pork production in the Moslem South.

Major changes may well occur in the Thai livestock economy in the near future. As the standard of living rises, there will be an increasing demand for meat, but at the same time mechanization, especially in the Central Plain, will reduce the need for work beasts, thus diminishing the present market for animals from the North East. It is thus important for North East Thailand to switch its emphasis from work cattle to beef cattle. Economic advisers have frequently stressed the potential of North East Thailand as a livestock region because edaphie and climatic factors rule out this area as an optimal region for rice production. Nor can it effectively compete with the Upper Central Plain in the cultivation of upland crops for export. "Breed and feed" is the aphorism that aptly sums up the crucial determinants of livestock development. The hardy Brahmin cattle should be upgraded by cross-breeding with imported beasts, local pasture should be improved where possible, and the upland given over to the production of fodder crops such as millet and cassava, which would stimulate the growth of a regional compound feed industry. Pig raising could comfortably fit in as a complementary sideline. If these changes do occur, then animal production in Thailand is going to show greater geographical regionalization than ever before.

REFERENCES

[1] *Thai Exports*. Thai Export Directory Ltd., Bangkok 1973.
[2] "A Ranch in Korat", *Investor*, Jan. 1971.
[3] Seminar on "Industrial Plant" held by the Agricultural Services Association, Thai Industry Association and ASRCT in 1970.
[4] *Agricultural Economy of Thailand*, U.S. Dept. of Agricultural Economic Research Service 1972.
[5] Khun Napaporn Lim, Advance Pharma Co. Ltd., Bangkok, personal communication.

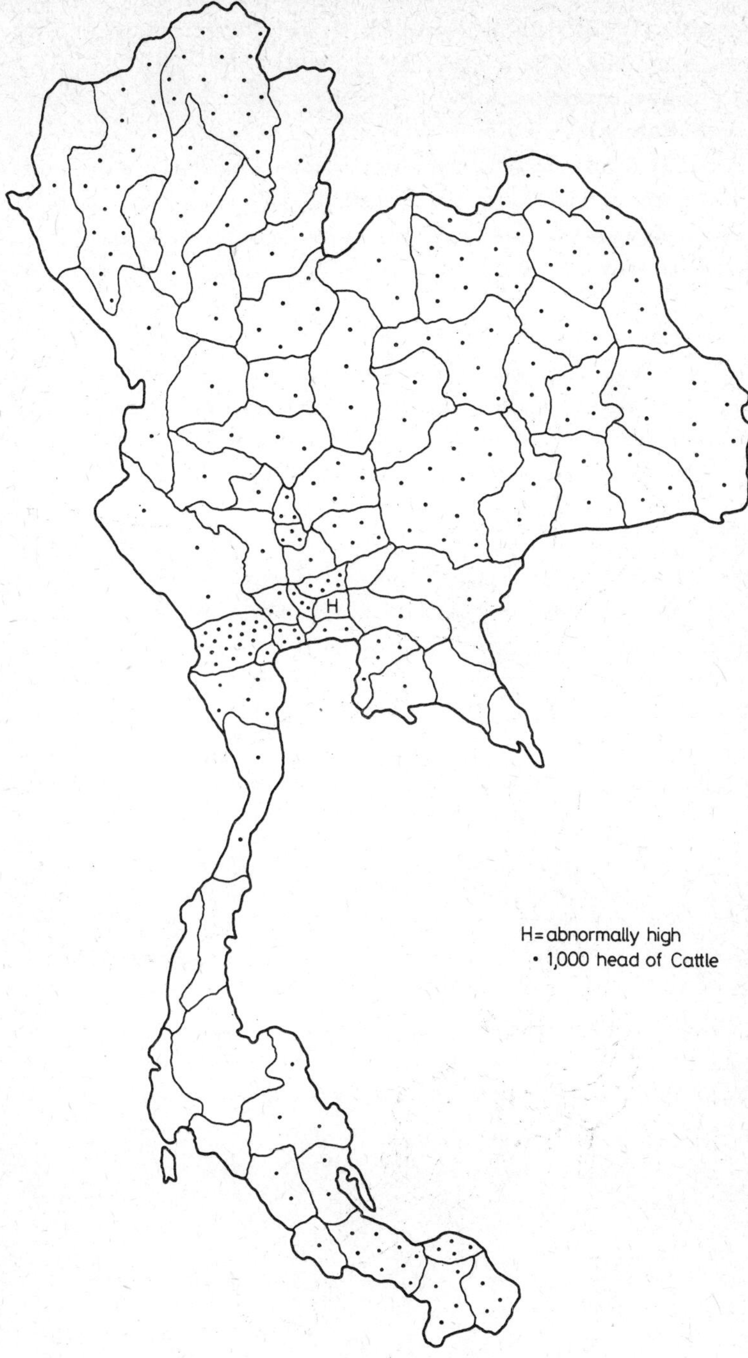

Map 24. Cattle slaughtered (Source: Livestock Report, 1970)

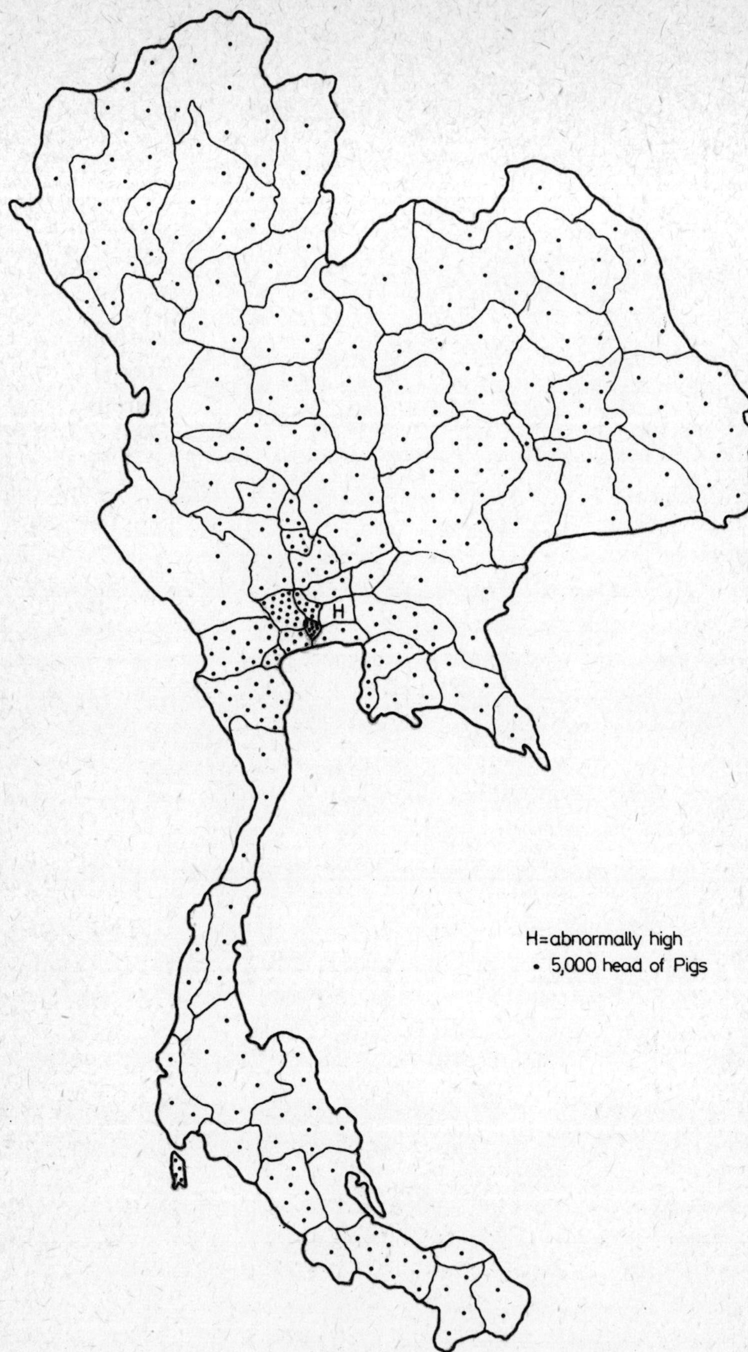

Map 25. Pigs slaughtered (Source: Livestock Report, 1970)

CHAPTER TWELVE

TRIBAL CULTIVATORS

No account of the agriculture of the country would be complete without reference to the tribal shifting cultivators of Northern Thailand, composed of a wide variety of tribal groups who are ethnically, linguistically and culturally markedly different from the Thai. However, these tribes are not enumerated for census purposes and official agricultural statistics take no cognizance of them. Nor does their produce enter the national economy with the one exception of opium, which though illegal manages to reach not only Thailand's addicts but a disturbingly large number of addicts in Hong Kong, London and New York.

A United Nations Survey Team reported that there were about 44,269 acres under opium in Northern Thailand [1]. in addition to which, the Hill Tribes cultivate a wide variety of subsistence crops including upland rice, maize, millet, sweet potatoes, yams, fruits, vegetables. They also keep many pigs and cattle which are allowed to run loose, while some tribes raise dogs for consumption. In addition, they may grow peppers for sale in local ethnic Thai markets or cellect for sale the leaves of the wild "miang" tea bush *(Camellia sinensis* var. *assamica)* which grows wild on the northern hills [2].

However, for many of the tribal peoples opium is the mainstay of their economy. There is a certain logic in its cultivation. It is a crop which has a peculiarly high value per unit of weight and volume and is not readily perishable. Hence it is economic to cultivate in the remote upland areas and then transport the long distances to the valleys on pack pony. Other crops such as rice and maize could not compete in lowland markets because of the high costs of production and transport facing the tribal cultivator. Moreover, many tribesmen are addicted to opium. For these reasons it is difficult to persuade or coerce them to give up its cultivation. Indeed, a fillip to Hill Tribe production was given by the ban on opium cultivation in Anatolia, Turkey. It is also doubtful whether opium cultivation could continue on the scale it does in Thailand without the assistance of various powerful and nefarious organizations such as the disbanded but still armed Kuomintang soldiers of the Burmese border, the Triad or Chinese Brotherhood (a secret society) and the European Mafia.

Map 26 shows the distribution of tribal shifting cultivators in Thailand, based upon various official sources [3], which must serve as an index to the land-use

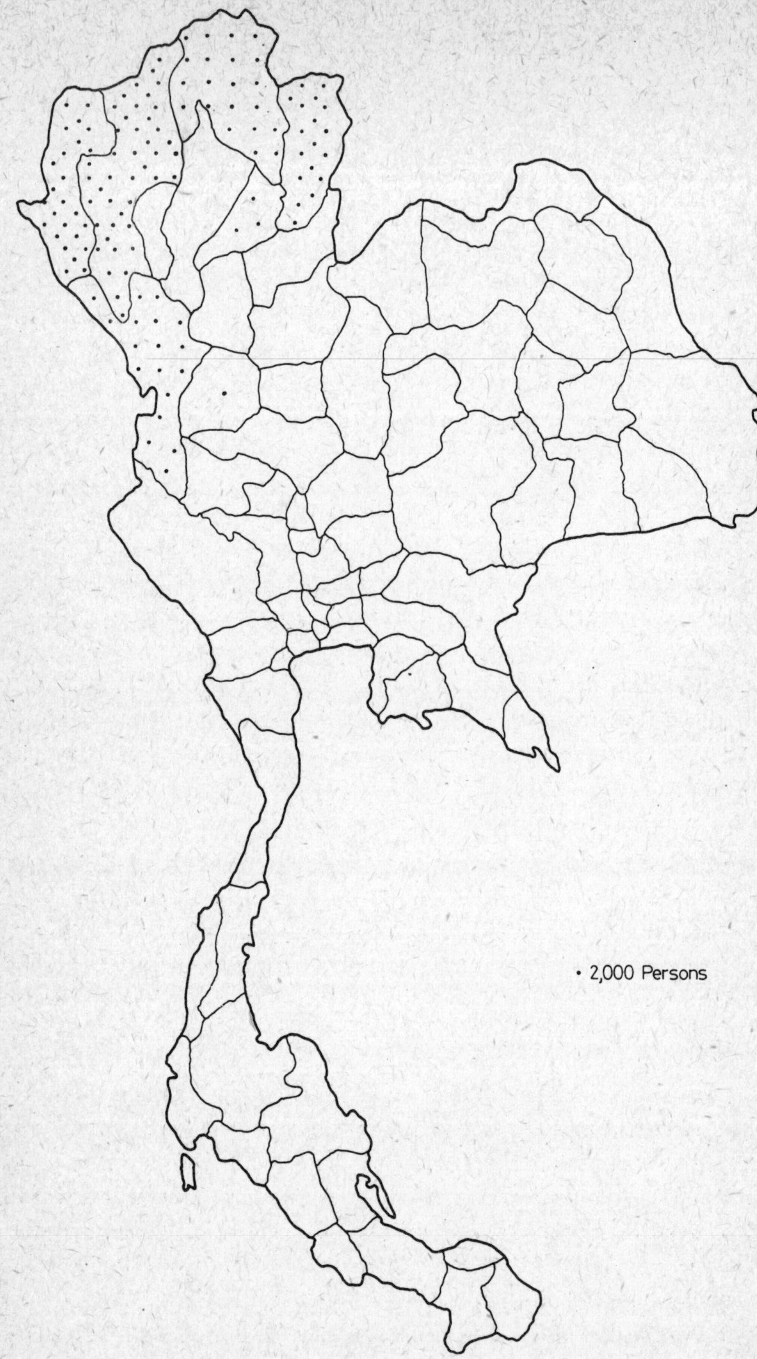

Map 26. Tribal shifting cultivators (Source: Various official sources)

under shifting cultivation. The affected provinces are Mae Hong Son, Chiang Mai, Chiang Rai, Nan, Phrae, Tak and Kamphaeng Phet, the highest concentration being in Chiang Mai. Tak and Kamphaeng Phet are not Northern provinces and there is some evidence that the tribes with their peculiar land-use are moving southwards into Kanchanaburi and perhaps Ratchaburi.

The lack of consistent, reliable data on Hill Tribe agricultural activities should not suggest that the tribal issue is marginal to Thai agriculture. On the contrary, it is one of great importance. Obviously, the cultivation of opium is a major problem not only for Thailand but also for the world as a whole, although it is probable that, government pressure notwithstanding, the Hill Tribes will persist in its cultivation unless they can adopt an equally profitable alternative cash crop. Secondly, the Hill Tribe population is increasing (perhaps more rapidly than the Thai) and pressure on the land is therefore also increasing. This leads to a reduction of the forest-fallow period in the shifting cultivation cycle with consequent destruction of the forest and its replacement by useless cogon grass *(Imperata cylindrica)*. Severe gully erosion results from the subsequent soil depletion and exhaustion. Not only does this endanger the livelihood of the tribes but it again brings them into conflict with the government which regards the forest reserves as one of the nation's most valuable resources. Whilst the integration of the Hill Tribes into the nation as a whole is a pressing political problem, agricultural research should be directed to lessening their dependence upon opium as a cash crop. COOP and SCHWASS broached the novel idea that sheep could be introduced to the Hill Tribes, thus providing meat and wool (for blankets) [4]. Sheep are not numerous in Thailand and at the time of the 1963 Census they numbered only 10,000 in the country as a whole and only 1,600 in the Northern Region. These sheep are mainly of local origin, resembling the more numerous Malayan and Indonesian breeds. COOP and SCHWASS consider that reliance should be placed on these tough local sheep rather than on the import, at great cost and risk, of exotic breeds from Europe or Australasia.

REFERENCES

[1] *Report of the United Nations Survey Team on the Economic and Social Needs of the Opium Producing Areas in Thailand*, 1967.
[2] Young, C., *The Hill Tribes of Northern Thailand*, Monogr. I, Siam Soc., Bangkok, 1962.
[3] Tribal Data Project, Tribal Research Centre, Hill Tribes Welfare Division, Public Welfare Dept., Ministry of Interior.
[4] Coop, I. E. and Schwass, R. H., *Report on the Possibilities of Introducing Sheep for the Hill Tribes of North Thailand*, New Zealand Embassy, Bangkok, 1970.

CHAPTER THIRTEEN

AGRICULTURAL TYPES AND REGIONS

The delimitation of agricultural types is an important aspect of any agricultural study. Analysis and development planning cannot be based on studying agricultural demands as if they were independent of each other and a good deal of attention is devoted in the agricultural literature to the delimitation of agricultural types and regions. The definition of agricultural regions in this study is based on agricultural properties, and an explanation of the regions so defined sought in physiographic, edaphic, social, economic and other external factors. Such external features are not therefore employed in the differentiation of types and regions.

The correlation between physiographic features and agriculture in Thailand is of a general nature only. There are variations in productivity between the highland areas of the north and east and the Kra Isthmus and the Central Lowlands. Similarly the highlands are characterized by more mixed agricultural regimes than the wet rice areas of the Central Lowlands. Natural conditions, however, are not the decisive factors in the formation of agricultural types and regions, but act rather as constraints on the available inputs of labour and capital. Rather it is the non-natural external conditions of technology and organization that are critical in determining the formation of agricultural types and regions. Thus, under the same physical conditions, a number of different agricultural regions may exist: witness the variation within the Kra Isthmus from mixed fruit and vegetables in the north to rubber and non-glutinous rice in the south.

The basic unit of any agricultural typology is, naturally, the individual holding. The analysis of Thailand's agricultural types and regions is, however, based on the province, since separate statistical data are not always available for individual holdings.

Among the possible bases for classifying agriculture in Thailand are size of holding, animal or crop combinations, and the presence or absence of agricultural innovations, but whatever the basis, or principle of division is chosen, it must be applicable throughout the statistical population that is to be divided.

Holding size is of little significance, since farming is characterized almost throughout by smallholdings. Large holdings are significant only in the provinces of Nakhon Nayok, Pathum Thani and Chachoengsao. Similarly, adoption of agricultural innovations, such as land improvement, tractors or fertiliser, is concentrated in a few areas, largely in the Central Plain. In addition, most farmers keep animals (buffaloes as draught animals, chickens and pigs), but although there are again concentrations in a few areas, mostly in the North East and Central Plain, they are not of major significance.

The most significant basis for a classification of agriculture in Thailand then is crop pattern, and crops therefore form the principle of division for the agricultural typology used in this study. Where holding size, the adoption of innovations or animals are important, these are indicated as sub-types of the main classification system. In order to provide an integrated understanding of the cropping pattern, it has been necessary to devise a method of classification based on combinations of crops rather than individual crops, since, in Thailand, only rarely does a single crop assume a position of dominance in any one area. Characteristically, crops are grown in combinations or associations, and any attempt to understand the patterns of cropland use must eventually move up to this level of description and analysis.

Such general diversity of cropping patterns led J. C. WEAVER to develop a statistical approach to the classification of crop combination regions—an approach which had been adopted in this study. The method is of particular value, since it is designed to deal with large areas rather than local studies [1].Although there are a number of definitions that might be employed to determine the identity of critical crop combinations, such as relative land occupancy or comparative production volume or value, the approach of using crop acreage has been adopted here.

An examination of the agricultural census reveals that the most significant crop combinations are composed of varying associations derived from the following crops: non-glutinous rice; glutinous rice; upland rice; maize; rubber; kenaf; mung beans; cassava; sugar cane; new cotton; soya bean; bird peppers; pineapples; chilli; castor beans; garlic; ginger; and groundnuts. A standard curve was therefore needed against which variations could be measured, but since the diversity of Thai agriculture effectively prevents the selection of an "average province" against which deviations can be measured a standard curve had to be defined on an abstract basis. Following Weaver the following theoretical curve was adopted:

monoculture = *100·00% of total harvested cropland in one crop*

2-crop combination = 50·00% in each of two crops
3-crop combination = 33·33% in each of three crops
4-crop combination = 25·00% in each of four crops

$$5\text{-crop combination} = 20.00\% \text{ in each of five crops}$$
$$6\text{-crop combination} = 16.67\% \text{ in each of six crops}$$
$$7\text{-crop combination} = 14.29\% \text{ in each of seven crops}$$
$$8\text{-crop combination} = 12.50\% \text{ in each of eight crops}$$
$$9\text{-crop combination} = 11.11\% \text{ in each of nine crops}$$
$$10\text{-crop combination} = 10.00\% \text{ in each of ten crops}$$

Using this as the standard curve, it is a straightforward matter to measure standard deviations from the theoretical curve as follows:

$$\sigma = \sqrt{\frac{\Sigma d^2}{n}}$$

where d is the difference between the *actual* percentage of land occupied by any given crop and the appropriate *theoretical* percentage in the standard curve, and n is the number of crops in a given combination.

The crop combination showing the smallest deviation from the standard curve was recorded for each province. Since the *rank* rather than the actual *magnitude* of the deviation was for this purpose, the simplified variant

$$\sigma = \frac{\Sigma d^2}{n}$$

was adopted.

For each province, all crops occupying more than 1% of the land were included. Deviations from the standard curve were derived for each crop combination to determine the lowest deviation, which then established the appropriate agricultural type for each region.

The agricultural typology of Thailand is therefore based on the following features:

A. Crop Combinations

The area occupied by each different crop. This forms the basic principle of division and the basis on which the agricultural regions are defined.

B. Organizational and Technical Features

Various sub-types are recognized as subdivisions of the main crop region classes, based on

1. the use of fertilizer,
2. the use of tractors,
3. the presence of shifting cultivation.

Where animals, large farms and improved land are regionally important, these are noted on the map, but do not constitute discrete sub-types.

Such a typology cannot be considered the only method of classifying Thai agriculture. Crop rotations and the degree of commercialization are possible alternatives, but reliable statistics are not available for these indicators. Under the system adopted, however, it is possible to distinguish 13 different regional types as follows:

Type 1 Non-glutinous rice. Generally this is one of the most productive and commercialized forms of agriculture. Here may be found an intensive use of fertilizers and tractors, and presence of a number of large holdings. The use of chemical fertilizers is also fairly widespread, though confined to less than 15% of the total number of holdings.

Type 1 × Non-glutinous rice with fertilizer and tractor use. Here, chemical fertilizers are used on more than 15% of the total number of holdings in each province. In the highest case, in Nonthaburi Province, 61·0% of all holdings use chemical fertilizer. Tractors are commonly found on over 20% of all holdings. Large holdings are also associated with this type.

Type 2 Glutinous rice. This is generally associated with low levels of mechanization and improvement. In only one area (Ubon Ratchathani) is it accompained by significant numbers of animals.

Type 2 × Glutinous rice with shifting cultivation. This sub-type contains the lowest levels of technology and farm improvement. Only 2·8% of holdings use chemical fertilizer and only 1·3% possess a tractor. The majority of holdings are small in size and there is a good deal of subsistence agriculture — although in certain cases opium is grown as a cash crop.

Type 3 Glutinous and non-glutinous rice. This type is also associated with low levels of organization and technology. No holdings are recorded in the agricultural census as using tractors, and only 0·3% as using chemical fertilizers. Holdings are mainly of intermediate size.

Type 4 Rubber. This is a cash crop grown generally on small land holdings. Levels of chemical-fertilizer use are under 15% and tractors are used on less than 20% of holdings.

Type 5 Rubber and non-glutinous rice. This is similar in technology and organization to Type 4, but is occasionally associated with the cultivation of chillies or the keeping animals.

Type 6 Mixed fruit and vegetables. Also similar to Types 4 and 5.

Type 7 Glutinous rice and maize. Fertilizers are used on less than 1·2% of all holdings and tractors on less than 4·5%. Animals are insignificant numerically —

but locally important on the Phetchabun range where cattle ranching is practised.

Type 7 × Glutinous rice and maize with shifting cultivation. This is a sub-type of Type 7. 3·0% of holdings have tractors and 5·8% use chemical fertilizers.

Type 8 Glutinous rice and kenaf. Few holdings are improved or mechanized — commonly less than 1%. This is a type similar to Type 2, only with significant areas devoted to kenaf. Non-glutinous rice is locally important.

Type 9 Mixed cropping. This is similar to Type 8, but contains also non-glutinous rice, maize, cassava, castor beans and groundnuts, with a preponderance of intermediate-sized holdings.

Type 10 Non-glutinous rice and maize. Maximum levels of fertilizer use on 6·7% of holdings, but tractor use is as high as 18·0%

Sub-Type 10 Non-glutinous rice and maize with high tractor use. Although fertilizers are used on only 3·8% of holdings, tractors are found on 23·7%, making this a more highly mechanized type than Type 10. Large areas are also under land-improvement co-operatives.

Type 11 Glutinous/non-glutinous rice with soya beans and garlic. This is more accurately a sub-type, since it is also characterized by significant levels of shifting cultivation. Animals and groundnuts vary regionally, while levels of technology and development are similar to Type 2 × .

Type 12 Non-glutinous rice and maize. A number of subsidiary crops are locally important: pineapples, upland rice, glutinous rice, mung beans and sugar. Technology and development are similar to Type 7.

Map 27. Agricultural regions of Thailand

Regions and sub-regions: 1 = Region 1: Non-glutinous rice; 2 = Sub-region 1 × : Non-glutinous rice with high fertilizer and tractor use; 3 = Region 2: Glutinous rice; 4 = Sub-region 2 × : Glutinous rice with shifting cultivation; 5 = Region 3: Glutinous and non-glutinous rice; 6 = Region 4: Rubber; 7 = Region 5: Rubber and non-glutinous rice; 8 = Region 6: Mixed fruit and vegetables; 9 = Region 7: Glutinous rice and maize; 10 = Sub-region 7 × : Glutinous rice and maize with shifting cultivation; 11 = Region 8: Glutinous rice and kenaf; 12 = Region 9: Mixed cropping; 13 = Region 10: Non-glutinous rice and maize; 14 = Sub-region 10 × : Non-glutinous rice and maize with high tractor and fertilizer use; 15 = Sub-region 11 × : Glutinous/non-glutinous rice, soya beans, garlic with shifting cultivation; 16 = Region 12: Non-glutinous rice and maize; 17 = Sub-region 12 × : Non-glutinous rice and maize with shifting cultivation; 18 = Region 13: Cassava and non-glutinous rice.

Groups: (a) = animals; (b) = large farms; (c) = improved land; (i) = pineapple, bird pepper, sugar, non-glutinous rice; (ii) = non-glutinous rice, cassava, rubber, upland rice, pineapple; (iii) = pineapple, upland rice; (iv) = glutinous rice, mung bean, sugar; (v) = non-glutinous rice, soya bean, new cotton; (vi) = mung bean, glutinous rice, soya bean, pineapple, chilli; (vii) = mung bean; 1: non-glutinous rice; 2: upland rice; 3: upland rice and new cotton; I = chilli; Pn = peanut; NG = non-glutinous rice; S = sugar cane.

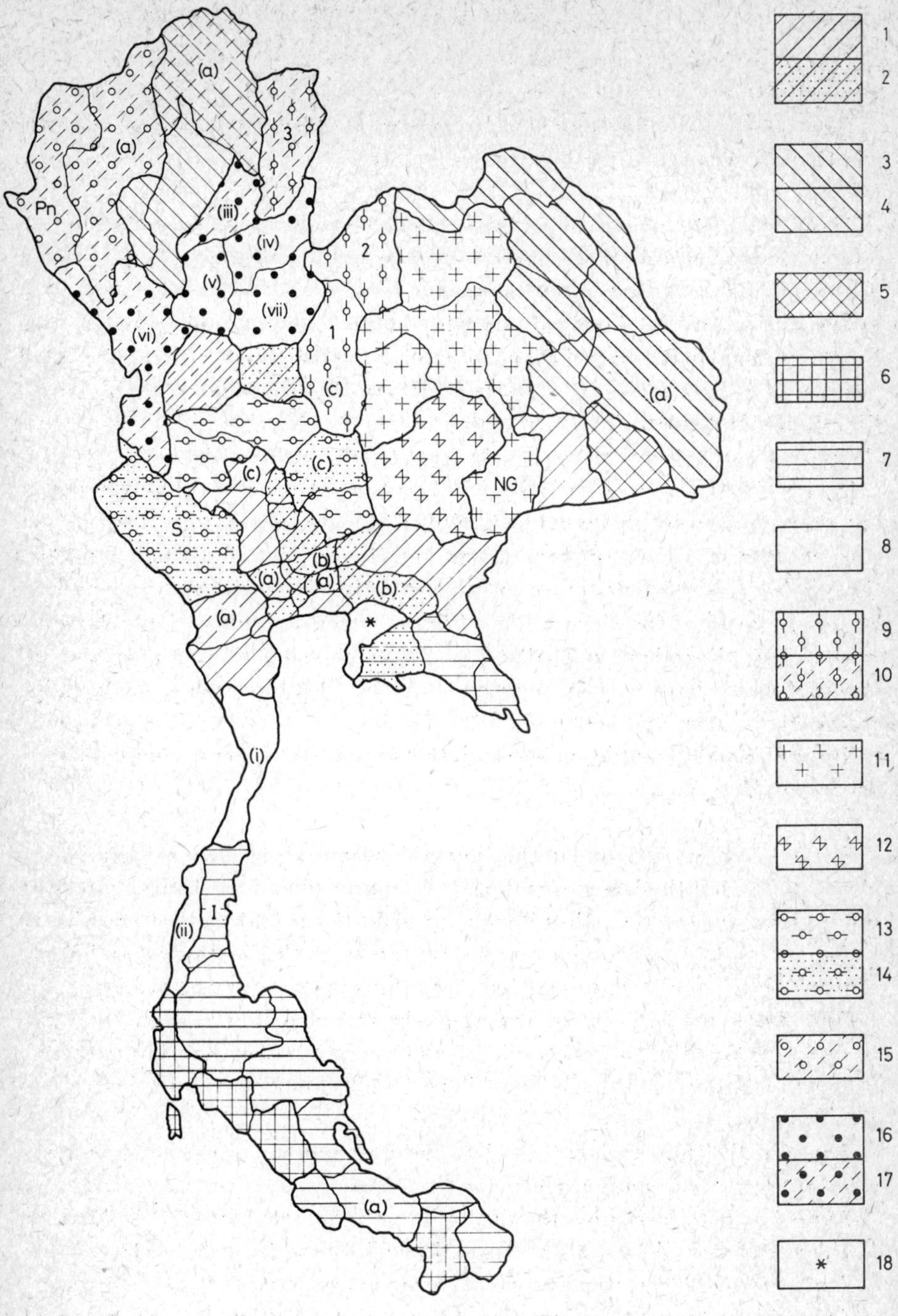
(a)
(a)
Pn
3
(iii)
(iv)
(v)
(vi)
(vii)
2
1
(c)
(c)
(c)
(a)
S
(b)
(a)
(a)
(b)
(a)
NG
*
(i)
(ii)
I
(a)
1
2
3
4
5
6
7
8
9
10
11
12
13
14
15
16
17
18
*

Type 12 × Non-glutinous rice and maize with shifting cultivation. This is a more subsistence version of Type 12.

Type 13 Cassava and non-glutinous rice. This is similar to Type 1 with the addition of cassava as a cash crop.

The above types give rise to a highly complex regional pattern, and since they do not always coincide with convenient physiographic regions, the regions are denoted throughout by the type classes.

The agriculture of Thailand may be subdivided geographically into the Northern Highlands, Central Plain, North East and Isthmus of Kra. Variety in terms of crops grown is greatest in the Northern Highlands, whereas the Central Plain is dominated largely by non-glutinous rice. The North East is in many ways the poorest agricultural area with the least fertile soils and the lowest levels of irrigation. The Isthmus of Kra is in some ways an extension of the northern area, but characterized by the growing of rubber in many areas.

In most areas, agriculture revolves around rice production, though in parts of the North and North East this means glutinous rice. The major rice growing areas are in the basin of the Menam Chao Phraya, where its cultivation is largely a monoculture. Elsewhere in Thailand, rice is combined with a number of crops, many of which have been introduced since the 1950s, as cash crops including maize, groundnuts and kenaf.

The regional differentiation based on the classification of crop combinations is shown on *Map 27*.

Types 1 and 1 × are found in the alluvial lowlands of the Menam Chao Phraya closest to its mouth. The most intensively improved and mechanized areas are grouped in a ring to the north of Bangkok and northwards along the Chao Phraya.

Types 2 and 2 × are found in two discrete areas: in the North East and North. Shifting cultivation and animals are important in the province of Chiang Rai.

Type 3 is found adjacent to Type 2 in the province of Si Sa Ket.

Type 4 is found in two areas on the southern side of the Kra Isthmus.

Type 5 is a more mixed version of Type 4 and is found on the northern side of Kra Isthmus.

Type 6 is the third region of the Kra Isthmus and lies on the north and west sides of the rubber and non-glutinous rice areas.

Types 7 and 7 × form a North-South belt in the highland areas of the Northern Highlands and the Phetchabun range, divided by the Laotian border.

Type 8 forms a transitional zone between Types 7 and 2.

Type 9 lies to the south of Type 8 in the province of Nakhon Ratchasima and to the north of the rice growing areas of the Lower Menam Chao Phraya (Type 1).

92

Types 10 and 10 × lie to the north of Types 1 and 1 × and provide a zone of transition between the rice growing areas of the Chao Phraya and the more mixed agricultural regions of the north.

Type 11 × is found in the extreme north of Thailand in the mountains along the Burmese border.

Types 12 and 12 × lie generally at a lower altitude and to the south of Type 11 × and contain a greater variety of crop types.

Type 13 is found only in Chon Buri province, although the cassava growing areas are also to be found in the neighbouring province of Rayong (Type 5).

Agricultural regions are dynamic and not static entities. The causes of regional variations are varied and complex, and a detailed analysis of the pattern of Thai agriculture awaits further study. Such a study should seek explanations not only in terms of physiographic, edaphic and climatic factors, however important these may be, but should also examine such factors as the role of religion, the social and economic factors governing the adoption and spread of innovations and the impact of external multinational aid programmes.

Existing studies of traditional agriculture tend to be based on European and North American paradigms and conclusions drawn from these areas are often applied without modification to the traditional world. This not only helps to account for the unsatisfactory nature of many explanations of agricultural patterns but also for the paucity of usable sources of information for such a task. The delineation of agricultural types and regions, however, provides an essential basis for this task of explanation, and also forms a base against which changes can be measured. Future studies, based on detailed work in the field, should concentrate on filling in the gaps in our understanding of such types and regions.

REFERENCES

[1] Weaver, J. C. "Crop combination regions in the Middle West", *Georg. Rev.*, 1954.

CHAPTER FOURTEEN

PROSPECTS AND POTENTIALS

It is difficult to predict the future for Thailand's agriculture with any certainty. The Thai proverb succinctly states: "Let the predicting doctor compete with the guessing doctor". In a very real sense the agriculture of any particular country cannot be considered *in vacuo*, but must be related to the prevailing economic and political climate of the world as a whole. Mainland South East Asia remains an area of potential or actual instability and conflict; at the same time few political scientists would be so rash as to predict Thailand's political future. In economic terms the entry of the People's Republic of China into the world market as a major exporter of agricultural produce, especially oilseeds and coarse grains, could have serious repercussions on Thailand's agriculture.

Nevertheless, certain statements may tentatively be made. Thailand will in the foreseeable future continue to be a major producer and exporter of rice, an activity for which the land is naturally endowed and the populace traditionally skilled. But changes may occur in the rice economy. Thus we suggest there may be a decline in glutinous rice cultivation both absolutely and relatively. Not only has this crop a very restricted export potential, but it is also possible that actual consumption of this starchy and heavy foodstuff will decline as the standard of living of the rural population increases and as the proportion of urban dwellers increases. Non-glutinous rice cultivation may then achieve a greater geographical concentration in the areas optimally suited for it. However, it is doubtful if the time has yet come when Thai farmers in any region will be prepared to abandon rice cultivation to any significant extent and rely upon the market place rather than their own farms for their sustenance. On the world scene it is to be hoped that Thailand, whilst retaining its traditional Asian markets, will be able to increase its share of the growing market for rice in North America, Europe and even Africa.

The future for upland crops is less certain. As far as rubber is concerned, it may be said that the course of natural rubber prices in the post-war period has shown its traditional pattern of sharp oscillations, but since 1960 the trend has been downwards. Competition from synthetic rubber, which will doubtless increase, has exacerbated this trend, and any significant expansion in Thailand's rubber production is therefore unlikely.

Although Thailand is by far the largest producer of kenaf and virtually the only exporter, the market does not seem secure. The increased production in 1966/67 placed a severe strain on handling capacity with the result that grades deteriorated badly. Problems are also created by the increased use of multi-walled paper bags, plastic bags and the development of polypropylene. It is, however, possible that the expansion of agriculture and industry in the Third World will produce an upsurge in the demand for sacks, whilst research in Thailand suggests the use of kenaf in the pulp industry.

Expansion in maize production for the Japanese market is likely, although Thailand has experienced difficulties in the maize export trade, arising over the inability to predict accurately export availability. In 1971 a contract for the supply of sorghum to Japan was obtained, but at present production in Thailand is plagued by internal marketing difficulties and pest attacks, whilst in the Asian market severe competition comes from Australian producers. It should also be noted that the writers' study of imports of agricultural produce into Japan in the last two decades reveals that she is unique not only in the wide fluctuations of the total imports of a particular agricultural commodity, but also in her rapid switch-over from one exporting country to another.

There is potential for the expansion of oilseed production in Thailand, such as castor beans, sesame and soya beans, all crops which in addition to their industrial use have a subsidiary use as livestock feed. The yields of sesame do not, however, compare favourably with the other two crops. Groundnuts are also a multi-purpose crop, providing human food (especially popular in the Far East), cooking oil and livestock feed, and although Thailand is not important on a world scale it is important in the Far Eastern market. Note that whilst in 1956 Thailand provided 56% of Japan's groundnut imports, in 1966 the figure had fallen to 0·9% as Japan turned to new sources in South Africa and People's Republic of China. This fact further underlines the precarious nature of Thai–Japanese trade and the impossibility of predicting the prospects for individual crops. To withstand market vagaries it is advisable that Thailand divides its crop between export and domestic livestock feedstuff.

Cassava has achieved remarkable success in South East Thailand and appears to be expanding rapidly in North East Thailand. Tapioca exports are at present tied to the European Economic Community, although Belgium in 1972 refused further shipments from Thailand because of inferior quality. (Quality control is a problem which besets many agricultural exports and needs to be tackled firmly.) The demand for cassava depends very much upon the price of cassava *vis-à-vis* grain. Moreover, since cassava is inherently inferior to grain, it must be supplemented by protein, and an increase in the world price of protein makes cassava a less attractive proposition. Also Thai cassava will suffer severely, if the

Common Agricultural Policy of the E.E.C. is scrapped. Likewise the European preference is for lean meat, whilst cassava produces fatty meat in the beasts that consume it. Lastly, the People's Republic of China has begun to export this commodity and this could transform the world situation, probably to Thailand's detriment.

It is hoped that there will be an expansion in the planting of fruit and vegetables for processing and canning, although optimal areas are geographically restricted. There is, however, great potential if Thailand can break through into the European and North American markets in the face of severe competition from the United States, South Africa and Australia.

In conclusion, it seems prudent to suggest that Thailand should attempt to diversify her agriculture. Thus upland crops should be produced not only for export but also for domestic livestock feed. The stalks and residues from sorghum, maize, cassava and rice can be fed to cattle and pigs, whilst compound feed industries could be developed based upon tapioca, maize, rice bran, fish and other basic materials. Many farmers, especially in the North East, would have to specialize in livestock production to an extent hitherto unprecedented and at a time when the demand for work beasts in Thailand is decreasing, consequent upon the mechanization of agriculture in the Central Plain. It would also be necessary to upgrade pastures by the planting of improved species of grasses. Although the government is well aware of the desirability of promoting a livestock industry, a rice farmer cannot be transformed into a cattle herder overnight.

What will all these changes mean for the agricultural regions of Thailand? They will doubtless mean increasing regional specialization, so that future maps will show more varied and more precisely defined agricultural regions. Certainly, the Upper Central Plain will continue as the most diversified agricultural area, but one can note the increasing diversification of the North East also.

The writers, having set down on paper what they at present know or believe, await with great interest the publication of future statistics and the opportunities for new field work which will enable them to continue the fascinating study of Thailand's agriculture.

LIST OF MAPS

LIST OF TABLES